# Passive Fear: Alternative to Fight or Flight

# Passive Fear: Alternative to Fight or Flight

✦

## When frightened animals hide

E. Norbert Smith, Ph.D.

iUniverse, Inc.
New York  Lincoln  Shanghai

# Passive Fear: Alternative to Fight or Flight
## When frightened animals hide

iUniverse books may be ordered through booksellers or by contacting:

iUniverse
2021 Pine Lake Road, Suite 100
Lincoln, NE 68512
www.iuniverse.com
1-800-Authors (1-800-288-4677)

ISBN-13: 978-0-595-39096-0 (pbk)
ISBN-13: 978-0-595-67665-1 (cloth)
ISBN-13: 978-0-595-83485-3 (ebk)
ISBN-10: 0-595-39096-X (pbk)
ISBN-10: 0-595-67665-0 (cloth)
ISBN-10: 0-595-83485-X (ebk)

Printed in the United States of America

In fond memory of my grandparents, Hans and Emma Hansen, and my mother, Opal Smith. They instilled within me a lasting awe and respect for all things living, and for their Creator. That wonderment continues to deepen with a life devoted to studying animals great and small.

# Contents

# Foreword

*By Stanley Robertson, Ph.D., Southwestern Oklahoma State University.*

Just inside the north entrance to Nielsen Hall, the physics building of the University of Oklahoma, is a sculptured relief mural bearing the inscription "In nature's infinite book of secrecy, a little can I read." If this seems a modest statement of modest accomplishment, it nevertheless implies that it may take both talent and passion to learn anything new at all. As a young physicist, I had a rather chauvinistic view of the applicability of the comment. But in collaborations and discussions of science with Dr. E. Norbert Smith, Ph.D. over the past thirty years, I have seen that it takes both passion and talent to learn *anything* of nature. In *Passive Fear: Alternative to Fight or Flight,* Dr. Smith (hereafter my friend Norbert) gives a first person view of how one becomes driven to know and to relentlessly pursue knowledge of wild creatures.

To learn of nature requires the use of appropriate tools. For the study of animals in the wild, Norbert developed his own radio telemetry systems with transducers capable of encoding heart and respiration rates, body temperature, and other physiological variables. It was my privilege to assist with an analysis of some of his temperature measurements for alligators and turtles. Since physicists normally have their problems dealing with nonsymmetrical geometries, it should be no surprise that my first efforts at understanding the data were slightly less than useless. While I struggled with older data and the separation of physiological and physical effects, Norbert continued with new observations that extended his studies of an anomalous change of heart rate that he had first observed in frightened alligators.

This book is the account of these unexpected observations and how they turned part of accepted biological science on end. As seems to be often the case, the new findings shed some light on a human problem; in this case, the tragedy of sudden infant death syndrome (SIDS). One never knows where a good question will lead.

In addition to the work delineating the scope of the fear response and the physiological control of heart rates, this book is superbly entertaining. How many persons do you know who both could and would go to the trouble to teach wild swamp rabbits to dive in a laboratory in order to observe what happens to their heart rates in the process? Such endeavors were captivating for students assisting in the laboratory and this is part of why Norbert is the best motivator of students that I have ever known.

I won't spoil the reader's fun by a poor recapitulation of what is better said in the book. So by all means, just enjoy a first hand look at some excellent and entertaining science!

# Acknowledgments

The studies that made this book possible were dependent upon numerous people, institutions and granting agencies. My mother and grandparents instilled a life-long love and respect for living things and encouraged my natural curiosity. The late Professor Perry Jones of Southwestern Oklahoma State University introduced me to electronics and amateur (ham) radio while I was still in high school. Dr. Hobart Landreth, also from Southwestern, taught me the joy of animal research and the utility of radio telemetry for studying wild animals. Unfortunately he was killed in the prime of his life in a tragic canoeing accident.

Ted Joanen and Larry McNease of the Rockefeller Wildlife Refuge, Grand Chenier, Louisiana, introduced me to alligators. It was love at first sight and they opened many research doors for me and taught me many things. The late Dr. Clarence Cottam, director of the Welder Wildlife Refuge, provided financial support, Refuge access, encouragement, and years of friendship.

Numerous grants made the studies possible. Some of the pivotal grants included: Baylor University pilot grant, several National Institute of Health and National Science Foundation grants, Caesar Kliberg Foundation Graduate Fellowship and *National Geographic* support.

Many supportive professors helped design and conduct many of the following experiments. The late Dr. Gene Crowder and Dr. Fred Gehlbach of Baylor were particularly helpful. Drs. Fred White and George Bartholomew of the University of California at Los Angeles helped design many of the alligator thermoregulation studies and motivated me to a lifetime of field studies. Dr. Francis Rose of Texas Tech University extended to me time, advice, patience and friendship. And Dr. William Portnoy, also from Texas Tech, showed me the role of basic physics in attempting to understand the thermal characteristics of living alligators.

Many excellent students, some borrowed and some my own, made perhaps the greatest contribution to my success. It was they who got muddy and insect bitten and obtained most of the field data and animals. I thank each of you from the bottom of my heart. Two students in particular helped collect hundreds of wild alligators, without complaint. They were Gene Keepers and Steve Adams. Thanks guys.

To physicist and my dear friend Dr. Stan Robertson a special thanks. We first met at Fort Hays Kansas State University, where he become hooked on alligator thermoregulation…an addiction that remains after decades. It was he who taught me, "All science is multidisciplinary." His knowledge of thermal physics and math made me look wiser than I am. Thank you Stan.

My dream of publishing a book would never have become a reality without the tireless help of my two children Weena and Jayson. Weena teaches high school English and, in spite of a family and busy schedule, poured over untold revisions of the manuscript and translated my Oklahoma gibberish into a proper language. She still laughs at my uncanny skill of changing verb tense mid-sentence without warning and my seemingly random use of commas. Jayson taught me how to use and even feel comfortable on a computer and the importance of saving back up copies. Thanks kids. I owe you.

# Prelude

What does it mean to be afraid? How does fear change the workings of the body? Are there different kinds of fear? If so, does the body respond differently? What is the relationship between fear and illness?

This book is about the physiological and behavioral responses to one important aspect of fear. Fear is a strong emotion, but it is more. It alters behavior and affects virtually every part of the body. Fear changes our very chemistry, even our perception of things around us. The effects of intense fear may linger for hours or days. In man, chronic fear can cause physical and mental disorders. Animals lacking a well-developed fear response are at greater risk of predation.

For over three decades, I have studied the behavioral and physiological responses of wild animals to fear. Those studies have led to some startling and unexpected discoveries. Before we take up the alluring research trail, allow me to summarize briefly the traditional view of fear and its consequences on the body. With this brief summary, you will know as much about the classic fear response as I did when I embarked on my journey of discovery.

When a rabbit or other prey species becomes aware of the approach of a predator, it has two contrasting survival options. It can run, or it can hide. The first approach has been studied for over 70 years and was initially called the "fight or flight" response. Today it is more appropriately called the active defense response to fear. Wide ranging and complex physiological adjustments accompany it. The orchestration of the adjustments prepares the animal for prolonged strenuous activity. One of the trademarks of the active defense response is a marked acceleration of the heart.

The alternate behavioral response to fear is called "passive defense" and is just now beginning to be widely appreciated. This opposite response is just as important for an animal's survival as is the active defense response, and both have significant implications for man and beast. Passive defense is characterized by a marked slowing of the heart, and by the behavioral response of the cessation of activity by crouching or hiding and, in extreme cases, by feigning death. Our native opossum playing dead is a classic and well known example. This passive defense response is the major subject of this book.

Dr. Walter Cannon, of Harvard Medical School, was the first to systematically investigate the physiological response to fear. Starting in 1929, he clearly showed that fear produced dramatic physiological changes that helped animals survive during an emergency. In the laboratory, when an animal was confronted with situations that evoked pain, fear, or rage, a well-defined set of physiological reactions prepared it to meet the threat. Dr. Cannon was the first to use the term "fight or flight response" to describe the event, and every student of physiology is introduced to the concept.

Other investigators have continued to explore the complex interplay of physiological and behavioral responses during the active defense response. Some of the physiological adjustments described for birds and mammals include increases in activity, body temperature, heart rate and blood flow to skeletal muscle. Breathing and oxygen consumption increase. In the liver, glycogen stores are converted to sugar while blood flow to the skin and digestive systems are reduced. Most of these adjustments are controlled by the sympathetic portion of the autonomic nervous system and involve the release of adrenaline from the adrenal glands. Adrenaline intensifies and prolongs the responses. Together these alterations enhance prolonged strenuous activity, such as fighting or running, and increase the likelihood of survival under extreme circumstances.

Consider a typical example, such as the approach of a coyote toward a rabbit. If the rabbit runs, the coyote will most likely see it and give chase. The rabbit is literally running for its life, and its survival depends on speed, agility, and an instinctive knowledge of hiding places in the area. More often than not, the rabbit will make a sudden turn for cover and elude the predator. To the casual observer this may appear a trivial event; for the animals involved it is a matter of life and death. Success to the coyote means a meal and perhaps food for its young. Continued failure will result in starvation. To the rabbit, every predator-prey encounter is a life-threatening event. It only loses once! Death of a mother may also kill or place hardships on its young. The physiological adjustments associated with the active defense response help the rabbit survive many such encounters.

What if the animal is cornered or caught by the predator? Fear is transformed to defensive anger. The classic "fight" response ensues. Most mammals will aggressively defend themselves or their young. The most loving house cat arches its back, spits and becomes a formidable foe when threatened by an unfamiliar dog. Again, a struggle for survival occurs and the active defense response is involved. The same physiological responses that increase speed, agility and sensory acuity while fleeing now prepare the animal for a vigorous and sustained fight.

Man responds to life threatening situations in much the same way. It is the active defense response that accounts for the unusual strength, endurance and split second reactions observed under emergency conditions. Striking examples have been seen when a loved one is pinned under an object too heavy to be moved, or inside a burning house, during battlefield situations or other such extreme threats to survival.

At times, the active defense response is inappropriate for modern man. Prolonged fear, anxiety or anger without an appropriate releasing activity may lead to frustration and stress-related abnormalities. This is an inherent hazard for people living in modern industrial societies. Fear in our modern world results in the same readiness for physical activity as occurs with animals, but with one important difference. There may not be a socially acceptable release. An excellent example occurs when a police officer is called to a "domestic disturbance." He or she rushes to the scene, aware that officers are sometimes killed under similar conditions. The officer's life is on the line and the active defense response commences. But when the officer arrives, the dispute may be over, and no physical activity may be required. A suppressed frustration occurs. It is well documented that repeated, unexpressed anxiety, fear and anger can lead to stress-related illness. There is still much to be learned about both the active and passive defense responses to fear, especially as they relate to man.

With this brief introduction to the responses to fear, let's explore the passive defense response in detail. The trail begins in an unlikely place with a remarkable beast.

# 1

# Something is wrong!

*But now ask the beasts, and let them teach you;*
*And the birds of the heavens, and let them tell you.*
*Or speak to the earth, and let it teach you;*
*And let the fish of the sea declare to you.*

—King Solomon

During the second week of my first semester of classes at Baylor University, our Ecology class made a field trip to the Aransas Wildlife Refuge in south Texas where Whooping Cranes winter. It was a typical graduate-level life science class field trip: in the early morning hours, we were wading waist deep through the swamps collecting frogs and poisonous snakes to be sexed, measured and released the next day. In the distance, I saw the unmistakable eye-shine of an alligator across a small pond. The eye shine resembles two glowing embers dancing on the water. I had seen it before. My heart raced with excitement.

"Be quiet!" I announced, "I'm going to catch an alligator for the class."

No one saw the alligator and no one believed me. I was not surprised. After all, I was the newest graduate student, and from Oklahoma—a state not renowned for its alligator population. Undaunted, I called the alligator by imitating the grunting alarm call of young alligators and splashed my hand in the water. The alligator immediately swam midway across the pond toward me.

My professors and classmates saw it. For the first time, they became quiet and cooperative. With haste, they retreated to higher ground well behind me so I would have plenty of room to work. Again I called the alligator. It swam right to my feet and I jumped on it. It was a young male barely six feet long. I showed everyone several of the special adaptations alligators have that equip them for an amphibious way of life, including their protective third eyelid, closable external nares (nostrils) and closable ear flaps. After a female student tried unsuccessfully to put it to sleep by rubbing its belly and was nearly bitten, I released it unharmed and a wiser alligator for the experience. For me, the event was not only exhilarating, it was also immensely profitable.

Capturing that alligator on the Ecology field trip early in my graduate program earned the respect of my fellow students and convinced my professors I was a serious graduate student. Within a week, Dr. Gene Crowder, my major professor, applied for and obtained a Baylor University pilot grant to help me begin my study. Without that single fortuitous event, I may have floundered for months as a graduate student. I knew little about experimental design and even less about obtaining research funds. My professors at Baylor helped plan the alligator thermoregulation study using the telemetry system I had designed in Oklahoma the year before. They also helped secure research funding from the National Geographic Society and National Science Foundation. The future research at the Welder Wildlife Refuge would have been impossible without major financial support. Dr. Fred Gehlbach, another member of my graduate committee, had worked at the Welder Refuge and made arrangements for me to visit and meet the director.

The refuge is privately owned by the Rob & Bessie Welder Wildlife Foundation and is located in south Texas, about 35 miles north of Corpus Christi. It consists of 7,800 acres of coastal prairie and is bordered on the north by the meandering Aransas River. Large cattle ranches lie to the north and south. As a private refuge, it is not accessible to the public. Due to hunting restrictions and few visitors, wildlife on the refuge is relatively tame and expensive research equipment may be left unattended without fear of vandalism. There are over 100 white-tailed deer per square mile. Javelinas flourish. Coyote, bobcat, foxes, raccoons and armadillo are often seen crossing the refuge roads late at night. Jaguarondi, otter-sized tropical cats seldom seen in zoos, and mountain lions are spotted on occasion. Alligators abound.

That first visit to the Refuge was another pivotal event that forever changed my life. Being an impoverished graduate student, I rode my old 175 cc Honda motorcycle from Waco to south Texas. Keep in mind the refuge is private and the local public has little access or knowledge of what goes on there. I was bearded, road-weary, sweaty and less than two miles from the Welder Refuge entrance. And I was hopelessly lost. Choosing to ignore the ominous "Keep Out" and "Trespassers will be shot" signs, I turned into a large Texas cattle ranch to seek directions. Just like in the movies, cowboys on horses met me. In addition to the saddle and rider, each horse carried a bedroll, two water canteens, rope and a very long rifle. I explained my plight, and was promptly informed in no uncertain terms, "This road does not lead to no damn refuge."

As I re-started the motorcycle, the nearest and meanest looking cowboy accusingly added, "I was a work'in here more'in 12 years and never heard a no such place." With renewed motivation I re-traced my route, re-checked my map and shortly reached my destination. Adrenaline clears the mind.

My reception at the Welder Refuge headquarters was diametrically opposite the encounter with the ranch hands two miles away. Dr. Clarence Cottam, the Director, could not have met me with more enthusiasm had I been a Nobel Laureate wanting to film a major research documentary there. He made this lowly graduate student feel not only welcome, but that my proposed alligator research was the most important research ever to be done at the facility. He provided lodging in the spacious dormitory and, after I stowed my bag and had a long cool shower, he took me out to eat the best Mexican food I've ever had. He rearranged his schedule so he could spend the entire next day showing me the refuge and its facilities. It was as though he had been waiting his entire life, hoping someone would study the alligators. We toured the natural history museum, small but adequate library, study areas, zoology, botany and necropsy laboratories and animal holding facilities. Chain link fenced kennels complete with concrete floors and running water from a previous coyote study would hold alligators nicely.

Then we toured the refuge. Driven like an all-terrain military vehicle, Dr. Cottam's old "Gold Bomb" Chevy sedan provided transport. Although there are nearly 12 miles of paved roads, Dr. Cottam seemed averse to use them other than as landmarks. We visited every place he had seen or heard alligators, or where anyone else had seen or heard alligators, or perhaps thought they had seen or heard alligators. He drove the Gold Bomb like a madman through mud and trees and bamboo that would stop an elephant. We continued on and visited each alligator nest site that had been used for the past five years.

He proudly showed me boats and motors, and assured me I had complete access to everything on the refuge, and they would gladly buy anything else I might require. He genuinely made me feel that my proposed alligator research was significant. Years later, I found that he treated every student that way. Each student was made to feel his or her research was the most important work being done at the refuge. With that kind of motivation, how could anyone fail? Few did. During my last year on the refuge, I was one of 54 graduate students from all across the United States, Canada and beyond. What a place to study, and what a motivator Dr Cottam was.

Dr. Cottam also encouraged me to apply for a Welder Wildlife Foundation Fellowship. I applied and was awarded full support for my research, books and tuition at Baylor University (and, later, at the University of California, and finally at Texas Tech University.) That trip not only provided me with access to research alligators and an excellent place to study them, but also paid for my entire graduate education. Each of the Welder Fellows lost a friend when Dr. Clarence Cottam died. He instilled a love of nature and confidence in research like no one I've met before or since. I miss him still.

After successfully completing a preliminary telemetry study with alligators at the zoo in Waco, Texas, I returned a second time to Welder. That was a collecting trip and there was only enough time to go out one night. Although staff and resident students had heard about my catching the alligator on the Aransas Wildlife Refuge and my proposed study, none thought I could catch alligators here. They argued the alligators here had recent memories of poaching and were far too wary. Even Dr. Cottam hinted that I might need more than one night to secure my study animals. My success hung in the balance. If only I had more actual field experience with alligators, or an experienced assistant…

It was an ideal night: dark with no wind. I had an inexperienced student helper, but in less than an hour, we had caught three alligators, and were able to return to Baylor the next morning. Soon my classes were out for the summer, and I was anxious to study alligators under natural conditions. The two smaller alligators had been used in a preliminary study and would be released. The third was outfitted with a sophisticated radio telemetry device to record three different temperatures and heart rate. It, too, was about to be set free. It was an arduous six-hour drive from Baylor University to Welder, as I was concerned about the alligators getting too hot—so I stopped every couple of hours to wet them down.

Shortly after arriving at the refuge, I drove to Big Lake and released the two smaller animals. Next, I switched the large alligator's radio transmitter on for the last time. Everything appeared to be working properly. One final calibration check revealed the three telemetered temperatures and heart rate were indeed accurate. With great excitement and considerable trepidation I released the telemetered alligator into Big Lake. Once released, my chance of recovery was next to nil if the transmitter failed. Alligators are intelligent and become extremely wary after capture. The directional antenna provided a strong signal and everything was working well. As expected, as soon as the alligator submerged underwater its heart rate slowed dramatically.

As far back as 1929, Dr. Scholander of Norway found that seals slowed their heart rates when forced under the water. Numerous investigators confirmed the results and the term "diving bradycardia" was coined to describe the marked slowing of heart rate with submergence. Virtually every diving animal tested in the laboratory shows the response and every physiology textbook has illustrations describing it. (Typically, even the human heart rate slows 30 percent when the face is submerged in cold water.) It is coupled with an entire suite of responses, including reduced blood flow to skeletal muscles and digestive organs, resulting in significant savings of oxygen, and it helps explain how many diving animals can remain submerged for hours. Alligators had been timed and could hold their breaths for over two hours without drowning. Other investigators had tested alligators and caiman, and each time the animals were forced underwater, they exhibited a marked slowing of their heart rates. My new telemetry system simply confirmed that the alligator was doing what it was supposed to do when submerged. But I had little time to bask in the confirmation or to sleep.

After releasing the alligator, I spent most of the night building a blind near Big Lake. Directional antennas were erected and meteorological equipment set out. I was able to record water temperature at several depths, mud bank temperature and air temperature, as well as wind direction, wind velocity, humidity and the intensity of sunlight. The radio signal confirmed the presence of the telemetered alligator. It was sleeping underwater and probably had just its nose above water. There was nothing to do but wait for sunrise. Everything was ready. If only the alligator would cooperate…

Daybreak over Big Lake was spectacular. The eastern sky was reddening, and the dark glow on the water promised a new day. From my blind near the shore, I had a panoramic view of the 200-acre oxbow lake. A whiff of water lily fragrance infiltrated the musky aroma of the retreating night. Amidst a nearby bulrush thicket, the night chorus of green tree frogs yielded to the raucous arguments of boat-tailed grackles. To my right, a purple gallinule hen and her two chicks plodded noisily atop water lily pads in search of aquatic insects. Across the lake, the lonesome call of a solitary bullfrog greeted the approaching day. As the day brightened, cattle egrets left their communal nests in search of livestock herds. In the distance, the characteristic silhouettes of three turkey vultures gradually took form on a dead oak tree. It was a good day to be a zoologist.

Everything was in place and all the equipment seemed to be operating flawlessly. I was eager to learn whatever alligators would teach me that day. As a zoologist, I am a life-long student of nature, and, to me, an experiment is nothing more than asking nature a question. If the question is phrased correctly—that is, if the experiment is thoughtfully designed—the answer will be apparent. I was asking two important questions of alligators.

The first question was easy to understand and required little equipment:" How much do alligators in south Texas grow each year under natural conditions?" Growth data were available for numerous captive alligators and for wild alligators in Florida and southern Louisiana, but nothing had been published regarding the growth of unrestrained alligators living in Texas. To answer this question, I needed only capture as many alligators as possible, then weigh, mark, release and recapture them. This was the beginning of what I hoped would be a long-term study. The plan was to re-capture marked alligators year after year, plus new ones, and by keeping careful records, discover how much they grow, how much weight they gain and how far they travel in a year. The long-term study would provide funding and a defensible excuse to return year after year to this zoologist's paradise. (Zoologists sometimes need such excuses to explain to others why they do what they do.)

The second and more difficult question was "How do alligators regulate their body temperature under natural conditions?" Data abound on how other reptiles regulate their temperature both in the laboratory and under natural conditions. Nothing had been published, however, regarding thermoregulation of either captive or free-ranging alligators. This question would not be easy to answer, and required years of preparation and the development of a sophisticated radio telemetry system to monitor body temperature in three places along with heart rate. I was working toward a Master's degree in Biology at Baylor University, and alligator thermoregulation was the object of my curiosity and the topic of my Thesis.

Two large alligators, approximately 11 and 13 feet in length, appeared in open water at 10:05 AM. They were swimming slowly and floating high in the water. Within five minutes the telemetered alligator made its debut at the surface. (Observing non-telemetered alligators is crucial, because they serve as experimental controls. If the behavior of telemetered and non-telemetered animals is similar, it is evidence that the radio device has not altered the behavior of the experimental animal.) The telemetered alligator was about 50 yards offshore. It was approaching an old earthen dike that once served as a road. It swam slowly at the surface and was floating high, just like the non-telemetered animals. At the zoo I called this the "high floating posture" and discovered it was a form of basking used to increase body temperature. It was reassuring that wild alligators were exhibiting the same behavior I had witnessed with captive animals at the zoo.

Everything was working better than I had dreamed. Since alligators are cold blooded, the environmental temperature influences their body temperature. My alligator's heart rate appeared normal for the size and temperature of the alligator.

The temperature data was excellent and continued to confirm observations from the zoo.

It was difficult to comprehend. I was actually taking physiological measurements from a free-ranging alligator in its natural habitat. I was collecting field data of wild alligators. No one had ever done this. I was getting a glimpse of alligator biology no one else on earth had seen and it was thrilling! It was also exhilarating to know I designed the telemetry system that made the study possible. What a rush! Zoology is so exciting!

Suddenly the telemetered alligator submerged. There was no indication of fright. Perhaps it was diving for food. Everything was still working. No, it wasn't. Something was wrong. The alligator's heart rate did not slow when it submerged! Could the telemetry device have failed? All the equipment was working flawlessly yesterday and earlier this morning. There was no reason to doubt it. The signal sounded normal and the three temperature readings appeared accurate. The alligator's heart rate continued to beat at its normal rate throughout the 7.2-minute dive. This simply could not be right. It made no sense. It violated everything I had been taught as a budding physiologist. Alligators, along with most other diving species, are supposed to slow their heart during submergence. Every physiologist knows this. What could possibly be wrong?

My telemetered alligator had obviously NOT read the books. It was free-ranging in its natural habitat. Its behavior matched that of non-telemetered alligators. I knew it was healthy and the telemetry system was working. Everything checked out. It was just not conforming to the rules. Why not?

I remained out of sight in the blind for two days and nights and collected excellent data, but on the third night the alligator moved. For the next four days I literally lived in a 14 foot flat bottomed Jon-boat, following the alligator and recording data. I slept under a mosquito net and took measurements at two-hour intervals at night and every 15 minutes during the day. I desperately needed a shower and sleep in a real bed, but I couldn't leave. This was turning out better than my wildest fantasies. I was gaining information about alligators that no one else knew and it was thrilling. Field zoology research is at times difficult, but it is also exhilarating and immensely rewarding.

I collected hundreds of measurements from dozens of dives confirming my earlier observation that alligators dive without slowing their hearts. They simply fail to exhibit the classic diving bradycardia. The failure of the alligator to slow its heart when submerged haunted me day and night. It just did not fit into my concept of alligator biology.

At 2:12 AM the sixth night I lost the alligator. I was dead tired and half-asleep when the signal suddenly became weak. The telemetered alligator was moving! By the time I came to my senses and switched to the directional antenna, the alligator was out of radio range. Several days and nights were spent looking for the alligator to no avail. Sadly, I had time to contemplate my discovery.

Scientific research may be compared to putting together a puzzle, with some important differences. Neither you nor anyone else knows what the completed picture is supposed to look like. Nor do you ever have all the pieces to the puzzle. And more often than not, some of the pieces you are holding belong to a different puzzle. Research is most exciting when the results do NOT fit preconceived notions. As in this case, discoveries often serve only to unveil new mysteries.

Scientific discovery is like climbing a mountain that has never been climbed; one gets a glimpse of nature never before seen by anyone. That glimpse of nature is the enticement to find still higher mountains to climb. Discovering new questions to ask is as much a motivation in science as is finding answers to existing questions. Few non-scientists appreciate this aspect of science, yet it is a significant driving force. As with the fox, the chase is often more exciting than the kill. My mountaintop discovery that alligators diving under natural conditions do not slow their hearts clearly brought ever-taller mountains into view. I had to find out why.

I can not envision a more exhilarating craft. I shall attempt to share both the excitement and the quest, but first I must do some backtracking.

# 2
# Preparation for research

*Luck is a matter of preparation meeting opportunity.*

—Oprah Winfrey

Norbert on top of a large female alligator

One does not just wake up one morning, roll out of bed, and decide to investigate alligator thermoregulation at a wildlife refuge in Texas. Such a momentous affair requires years of preparation. Luck is also a necessary ingredient. A zoologist, like everyone else, sometimes needs to be at the right place at the right time. In retrospect, it seems most of my life had been spent preparing for that morning at Big Lake. Many seemingly unrelated events provided the background necessary for that study. A network of learned and willing people helped with experimental design, research funding and logistic support. If those people had not been available, or if certain events had not occurred at the right time, the study would not have come to fruition. Serendipitous events, as far back as high school and before, played key roles in preparing me for that morning.

Many of my earliest and fondest memories involve animals, especially reptiles and amphibians. High on my list of favorite childhood memories are toads—lots and lots of toads. Toads flourish in western Oklahoma. When injured, some species secrete a toxin strong enough to make a dog vomit and cause permanent blindness in man. Toxic secretion notwithstanding, toads are docile and quite safe to handle. The smallest child can apprehend them. Toads are friendly creatures and almost everyone likes them. They inhabit gardens and eat insect pests. During my preschool summers I spent endless hours siphoning water from the livestock-watering tank, making small puddles to attract toads. A nightly ritual, indelibly engraved in my memory, was practiced shortly after dark. My mother would get the "Big Flashlight" and we would count the number of toads attracted to my "toad ponds." Once there were 52! Toads, like other amphibians, do not require drinking water. They absorb water directly through their skin. The toads were attracted to the mud puddles to rehydrate. I still recall the elation I felt when, following a rain, a toad laid eggs in one of my toad ponds. Oh, the joys of childhood on the farm!

Hours were also spent playing with horned lizards, "mountain boomers" (actually called collared lizards, our Oklahoma State reptile), hog-nosed snakes and box turtles. Even as a boy, I was interested in the fear response of animals. I often placed a newly captured box turtle in the edge of the shade from a large elm tree. The wind created a mosaic of moving shadows that would frighten the turtle and cause it to pull into its shell and remain motionless for hours.

Hog-nosed snakes were particularly amusing. When teased, they would first puff out their necks and hiss (thus the nickname of "puff adder"). If touched roughly, they would roll over on their back and play dead. They were unresponsive on their backs, but if they were turned upright they would quickly roll over again. To play dead, a snake obviously must lie on its back! If the threat continued, hog-nosed snakes would then bleed from the mouth. (This is accomplished by rupturing fragile capillaries inside their mouth.) This simple trick more often than not fooled dog and boy. Horned lizards had a somewhat similar response. When attacked by a dog, they puffed up to reveal their thorny coat. If the pestering dog bit them, they responded by releasing a stream of blood from a sinus at the corner of their eyes.

Although collared lizards were difficult for a little boy to capture, I enjoyed frightening them and watching them scurry to the safety of their holes under a rock ledge. I remember peering into their lair, wondering what they were doing and what they were thinking while hiding from me. I found these responses enchanting even as a small child. Little did I know that much of my adult life would be spent studying fear responses in some of these same creatures.

Psychologists would consider these childhood events and conclude that they shaped my personality and interests, which later resulted in my becoming a zoologist. I disagree. I was already a zoologist and had even then discovered my research niche. All that was lacking was refinement of that same curiosity by education, and a place and equipment to study fear in wild animals.

My interest in electronics was blossoming along with my pursuit of biology. Early science fair projects included homemade electric motors, electromagnets, transformers and a crystal radio. In high school I obtained an amateur radio license and talked to ham radio operators in 23 states with a homemade radio transmitter. Even my novice call sign was prophetic: KN5PHD. I had a Ph.D. from the FCC! My favorite reading material was the "Amateur Scientist" portion of *Scientific American*. In the 9th grade I modified a project described in that magazine and was able to view the electrocardiogram of a water flea. (Water fleas are microscopic crustaceans that inhabit pond water.) It seems I have always been interested in the heart rates of animals.

I recognize in hindsight that this event was significant because it was the first time I used electronics and biology together. For years I considered the two pursuits unrelated. One of the reasons I did not attend college upon graduation from high school was an inability to select a major. I liked biology and electronics equally but was told the two were incompatible. Upon graduation from high school I joined the United States Air Force with a promise of electronics training. My early dabbling in electronics qualified me for Air Force electronics school.

Electronics school in the Air Force was challenging. In nine months we learned as much electronic theory as electrical engineers do in a four-year college program. Practical experience and the best-equipped teaching laboratories in the world compensated for our lack of math and physics. I had learned enough electronics in high school to whet my appetite, and graduated at the top of my electronics class in the Air Force. Soon I was repairing sophisticated Air Force communication equipment. During free time in the Air Force, I completed several correspondence courses in electronics. The last course I took was a course on color television. Taking that course was one of those chance coincidences that played a central part in preparing me for studying alligators at Big Lake. After an honorable discharge from the Air Force, I worked as an electronic assembly line technician with Texas Instruments. Late one night, I wandered around the plant and found the consumer applications laboratory. I asked some other technicians about it and they explained that the people there designed radio and TV circuits (known as consumer products). I applied for a job the next day and at the interview was asked several questions about color TV. My only exposure to color TV theory was that correspondence course two years before in the Air Force. That rudimentary knowledge of color TV got me the new position. Once again serendipity was at work preparing me for a date with an alligator.

The consumer applications lab was an exciting place to work. My responsibility, as an engineering technician, involved working closely with electronic design engineers. Soon I was given limited design responsibility and at home I started writing articles for *Popular Electronics* and other hobbyist electronic magazines. I made many important contacts at Texas Instruments—contacts that helped with the design of the first rattlesnake transmitters two years later.

Electronic design was fun, but something was lacking. I missed animals. At home, I converted an old refrigerator into an environmental chamber and was investigating hibernation in hamsters. I built a 100 gallon refrigerated aquarium and maintained saltwater animals, including an octopus and other animals I had collected along the beautiful coast of Maine.

I wanted to know more about animals and decided to attend college and major in zoology. My early childhood interest in animals was rekindled. Two weeks into college I met Dr. Hobart Landreth, who needed a radio transmitter to track rattlesnakes. My involvement in the rattlesnake telemetry project underlined the value of electronics in animal research. At the time the only radio transmitter available for field use only transmitted about 30 feet and lasted less than two weeks on batteries. Within two months I had designed a radio telemetry system that transmitted over 2 miles from inside a rattlesnake and lasted over 6 months on batteries. I graduated from Southwestern Oklahoma State University in three years with a degree in Biology, some valuable radio telemetry experience, and a newfound love of research. Two weeks with Ted Joanen and Larry McNease of the Rockefeller Wildlife Refuge in Grand Chenier, Louisiana, sparked a deep interest in alligators. I was eager to continue my education at Baylor University and begin my own studies of alligator thermoregulation.

One might ask, "Why study temperature regulation of a cold blooded animal?" Animals use two contrasting methods to deal with environmental temperature extremes. Warm blooded (or endothermic) birds and mammals are well insulated and generate adequate metabolic heat to maintain a stable body temperature. The arctic fox is equally active when the environmental temperature is -50 degrees or +70 degrees F. Such independence from environmental temperature is far from free. The cost of maintaining a stable body temperature is high, especially during cold weather. As the environmental temperature drops, metabolism must increase to provide heat. At very cold temperatures, the arctic fox must catch the arctic hare or starve.

In contrast, cold blooded (or ectothermic) animals are poorly insulated and depend on environmental heat to keep warm. If one places a lizard, fish, or snake in a refrigerator overnight its body temperature will drop to very nearly that of the refrigerator. It can not shiver to keep warm. Temperature affects biochemical reactions and thus all aspects of living organisms. There are many advantages in being warm for both groups of animals. Digestion and assimilation of food, nervous reflexes, muscle strength and speed and the sensitivity of sense organs all increase with warm temperatures. Even as a preschool child, I remember placing toads, lizards and small snakes in the refrigerator overnight and was always amazed at how slowly they moved the next morning. Even then I was interested in the effect of temperature on reptiles.

Under natural conditions reptiles (and, to a lesser extent, amphibians) use a combination of behavioral and physiological mechanisms to regulate their body temperature. When cold, they seek warm places in the environment. Basking in the sun is the most common method used by reptiles to increase their body temperature. When hot, they seek shade.

Reptiles also rely on physiological adjustments. For example, a basking lizard warms twice as fast as it cools due to increased blood flow in its skin while basking. Ectotherms live more economically than endotherms. As the temperature drops, ectotherms require less energy. Alligators at the Welder Refuge, for example, often eat their last meal for the year in November and do not eat again until March or April. They remain inactive and consume very little body fat. Hibernating mammals gain the same advantage by allowing their body temperatures to cool in winter.

The purpose of my alligator study was to determine the effect of behavioral thermoregulation on body temperature. Alligators were known to bask in the sun, and it seemed logical that this behavior increased body temperature. No direct data were available. Observations over one hundred years ago indicated that alligators crawled out of water to bask when the air temperature exceeded water temperature. That observation bothered me, because how could a wet alligator know when the dry bulb air temperature rose above water temperature? Monitoring three different body temperatures would provide an indication of the direction of heat flow. Heart rate might provide an indirect measure of blood flow and was expected to increase during basking. I would also monitor several environmental temperatures, wind velocity and relative humidity, plus a measure of the intensity of sunlight.

Alligators living in swamps often dig a large pit or "gator hole". During periods of drought these alligator holes often provide the only available drinking water for wildlife in the area, but the actual reason alligators dig these holes had long been debated. Perhaps close observation of alligators at the zoo and under more natural field conditions could illuminate these and other questions.

The first portion of the alligator telemetry study was to be completed at Cen-Tex Zoo in Waco, Texas. The zoo was an excellent place to begin. Although I only spent eleven days and two nights there, the telemetry system performed flawlessly and I made several significant discoveries. Alligators use behavior to regulate their body temperature, but the behavior is more complex and varied than anything described in the scientific literature. They are able to exert control over their body temperature by merely changing position in the water. At night, when the water was warmer than the air, they remained submerged. In the mornings, when the sun's warming rays reached the water, the alligators all swam to a sunny patch, inhaled deeply and floated high in the water. The sun soon warmed and dried their backs. Aquatic basking had not been reported for alligators, yet it resulted in a significant increase in body temperature. Heart rate also increased during warming, suggesting an increase in blood flow to the warm skin and muscle.

As described in the old scientific literature, alligators crawled out to bask when the air temperature rose above the water temperature. Observations at the zoo provided a simple answer to the old question of how a wet alligator can respond to the air temperature. The skin of a wet alligator in air would be cooled by the evaporation of water. Alligators at the zoo never went directly from deep water to bask on land. Instead, they always lined up in shallow water parallel to the shore with their backs exposed to the air. Only after their backs were dry did they crawl out to bask. They were responding to the dry air temperature and were not wet at all.

Basking ended when the alligator's body temperature approached 90 degrees Fahrenheit. They returned to the water for the remainder of the day. Water temperature in the small alligator pool reached 94 degrees on two extremely hot afternoons. This is only 6 degrees below the lethal temperature for alligators. Two different methods were used to avoid overheating. A modification of aquatic basking helped the alligators lose heat in the afternoon: they swam to shady portions of the pool and again floated high in the water. They periodically submerged to wet their backs. The evaporation of pool water helped them remain cool in a hot environment. This method is more energy efficient than the sweating or panting observed in warm-blooded species.

During the hottest part of the day, the alligators retreated to the cooler deep water at the bottom of the pool. I made another important discovery on two days when the afternoon heat reached near-record highs. The largest alligators would occasionally anchor themselves with their front feet and make hard scraping motions with one rear foot and then the other. A vigorous swishing of the tail immediately followed this unusual behavior. Later in the laboratory I confirmed hosing alligators down with warm water could trigger the same behavior. Clearly this behavior accounts for the alligator holes described by early naturalists. Alligator holes made by mature individuals often measure 12 feet long, 4 feet wide and 6 feet deep. In a small zoo in Texas, I discovered something that had puzzled naturalists for centuries. Alligator holes are dug as a simple behavioral response to high temperature. I would have missed this discovery had it not been for those two extremely hot afternoons. Research often depends on a measure of luck.

Another important discovery was made at the zoo and later confirmed at Big Lake. Previous published information indicated that an alligator's body temperature was always within one degree of the water temperature. At the zoo, the tele-metered alligator maintained its body temperature several degrees above the water temperature, even during the night. This was unexpected. At first, I doubted the accuracy of the telemetry measurement, but subsequent calibration confirmed the measurements. It required several years of intense research to elucidate the mechanism. Like mammals, large alligators produce sufficient metabolic heat to elevate their body temperature. I would not have noticed a significant temperature difference with smaller alligators. Heat loss to the water is reduced by a marked reduction in blood flow to skeletal muscle and skin. Why had others missed this important difference in field caught alligators? Their body temperature approaches the water temperature during the intense struggling associated with capture, because struggling increases blood flow to muscle and skin.

Time at the zoo passed too quickly. The days and nights spent living with alligators were profitable. Several important discoveries were made and the telemetry system was validated. The zoo provided an excellent beginning. I learned more during those eleven days at the zoo than could be learned with months of field-work. Access was easy and the animals were accustomed to humans. I was anxious to work with free-ranging alligators, but I knew the research would be more difficult at the wildlife refuge.

# 3

# Accidental discovery

*What reason, like the careful ant, draws laboriously together, the*
*wind of accident sometimes collects in a moment.*

—J. C. F. von Schiller

Two weeks elapsed after the telemetered alligator eluded me at Big Lake. Before losing it, I had collected thermoregulation data for 8 days and nights. Field data confirmed earlier zoo data at every point. Free-ranging alligators used aquatic basking to increase their temperature and basked when the air temperature rose above the lake temperature. They always aligned themselves parallel to the shore and allowed their backs to dry before crawling out and basking on land. Heart rate increased during basking, exactly as I had discovered at the zoo. I had learned much, but there remained a telemetered alligator to locate.

Several days were spent systemically searching Big Lake for the alligator. My assistant and I crisscrossed the lake numerous times with radio in hand, but heard nothing. We walked the entire shoreline twice to no avail. If it had gone to the river or to another lake it was unlikely I would ever find it. At its nearest point, the Aransas River is only a half-mile from Big Lake. I had other work to do and could not spend any more time looking for the telemetered alligator.

As we began collecting alligators for the growth-movement study, we started in the river, in hopes of finding the telemetered alligator. We carried the telemetry radio and kept it switched on, but we concentrated on catching alligators. For three nights we worked, marking and releasing alligators in the river. In addition to the growth-movement study, I was taking a population census of the number and size of alligators at the refuge. I intended to study population changes for the next several years.

The alligator population in the open marshlands of coastal Louisiana is tallied by counting nests from low flying aircraft. Since nesting females represent five percent of the alligator population, the total number of alligators in an area may be estimated by multiplying the number of nests by 20. In the dense vegetation of south Texas, the nests are difficult to spot by air or from the water. Instead, the population is estimated by taking night counts of alligators in an area. Our last night on the river was spent counting and sizing alligators. Estimating alligator size is relatively easy. For every inch between the eye and tip of the nose there is a foot of alligator. The telemetered alligator was not seen, nor its transmitter heard. Our work in the river was complete and we moved to Pollito Lake.

Pollito means "little chicken" in Spanish. It spite of its unimaginative name, Pollito is a picturesque lake—the essence of scenic calendars. Century old live oak trees surround it. Spanish moss does not range this far west, but other epiphytes festoon the eldest trees. The Welder Refuge is a bird lover's paradise, boasting over 500 resident and transient species. Scissor-tailed flycatchers nest high in isolated live oak trees and perform unending aerial acrobatics over the lake. Inca doves build nests in the low spreading branches and fill the morning with their winsome cooing. Hollow trees provide living quarters for black bellied whistling ducks. These strange goose-like ducks are game birds throughout Central America and migrate north to their summer breeding grounds in South Texas, where they are protected.

Cattails and bulrushes ring Pollito. Muskrats harvest the reeds and make distinctive feeding stations. Least bitterns fashion floating nests among bulrush thickets. Water shamrock, an unusual aquatic fern, competes with coastal Bermuda grass just beyond the shoreline. A luxuriant crop of water lilies grows in the shallows. Anhingas, appropriately called "snake birds," fish near the shore. Lacking oil glands, their feathers become wet and their body sinks. All that remains visible is their long snake-like neck with beak writhing back and forth among the aquatic vegetation. Late in the afternoons they fill a nearby dead tree with outstretched wings drying in the sun. On windless days muskrat and alligator trails, made the night before, form intricate geometric patterns in floating duck weed and mosquito fern. Field zoologists work in the grandest of environments.

Most of the graduate students studying at Welder worked by day. While collecting alligators, I owned the nights. South Texas summers are long, hot and muggy. Nights are less hot, the work unhurried, and the myriad night sounds are forever etched in my memory. Evenings were spent on the water in the midst of a symphony of nature. The playful call of coyotes was heard intermittently from dusk till dawn. Moonlit nights were filled with the ever-varied call characteristic of mockingbirds. The shoreline was alive with the clamor of frogs and crickets. The soft grunting of young alligators was heard near the bank. The plaintive call of whip-poor-wills and great horned owls provided a fitting tenor. The unmistakable shrill of a nearby whistling duck added the crescendo. In the early spring the thunderous bellow of bull alligators could be heard for several miles.

Most alligators are caught when the night is dark and windless. Moonlight illuminates the alligator hunter and the quarry retreat. Wind makes waves and the sound of waves breaking against the boat frightens the alligators and they submerge. During windy or moonlit nights we often caught fish or bullfrogs instead. (Perhaps I am part Cajun, because I prefer wetland game to store-bought meat.)

Pollito was populated with alligators of all sizes. Two females nested here, and judging by the pods of young alligators, both had successful hatchings the past two years. We collected over the entire lake for three nights and were beginning to recapture alligators caught earlier. This would be our last night on Pollito. We completed our final count and were about to head for the car. There, on the bank, about 20 yards from the lake was a large alligator, in its mouth a freshly caught raccoon. I had witnessed alligators catching snakes, fish and turtles, but this was the first time I saw one with a mammal. The raccoon appeared dead and the alligator was returning to the lake. What was that on its neck? Could it be…? Where is the radio? I quickly grabbed the telemetry receiver and switched it on…click, click, click. We had stumbled onto the telemetered alligator and the transmitter was still operating. If the alligator had been swimming in the lake, we would not have noticed the transmitter and we no longer bothered with trying to keep the radio operating. We quietly rejoiced. Lady luck remained my faithful consort.

We spent the remaining few hours of darkness constructing a makeshift blind on the point of land near the sighting. The directional antenna confirmed the experimental alligator remained near. From this location, I could cover nearly the entire lake. Anticipation of new data replaced the need for sleep—I recorded telemetry and environmental measurements at 15-minute intervals.

Low clouds muted the sunrise. An hour passed, then two. The telemetered alligator surfaced only to breathe. Triangulation pinpointed its location about 50 yards off shore, near a bulrush thicket. The thicket sheltered a least bittern nest, flagged for study. Behind the blind, a slight noise was heard. With the spotting scope I found the source. A white-tailed doe was nursing her twins. Pollito was serene. The telemetered alligator remained submerged with a normal heart rate of 25 to 35 beats per minute. Overcast skies delayed basking. No other alligators were seen on the surface.

A pickup truck was barely visible, parked across the lake to the east. It was unusual to see another graduate student. There were about 15 other students researching a variety of animals and plants at the refuge, but the area is large and most of them worked in the higher prairie or wooded areas. The pickup belonged to Will Regan. He used a canoe to investigate the hatching success of least bitterns and other shore birds. I certainly hoped he was NOT checking his nests that day! I didn't see him or the canoe. Maybe he was just fishing.

Another hour passed. Humidity covered Pollito like a wet blanket. The stillness was deafening. The telemetered alligator remained submerged with only occasional movement to reach the surface for breathing. Heart rate was 32 beats per minute and the body temperature was 5 degrees warmer than the lake. Zoo measurements were again confirmed. The clouds were thinning. Maybe I would soon get some thermoregulation data. Anticipation quickened my own heart rate.

Abruptly, Will's canoe emerged from a point of land across the lake. He was not fishing. His German shepherd dog, Napoleon, was with him. Maybe he would see my blind and stay away. It would be tragic to frighten the alligator again after only three hours of lackluster measurements. Will slowly zigzagged toward the blind. He stopped among the reed thickets at each flagged bittern nest to record the number of fledglings and condition of the nest. He photographed each flagged nest, and also discovered a new one. He marked the nest with a flag and took several photographs. He continued has census. The alligator's heart rate remained at 32 beats per minute, but Will was coming closer. I tried to wave him away. He returned the wave. I was near panic; he thought I was just being friendly. He was heading toward the marked least bittern nest, ten yards from the submerged alligator.

Suddenly the alligator's heart nearly stopped! Without warning, it dropped from 32 beats per minute to less than two. The radio signal betrayed the location of the alligator. It did not move as the canoe passed overhead. Had the transmitter malfunctioned? No, the signal remained strong and the temperature measurements were consistent. Only the heart rate changed. Will continued his tally and began to work his way toward the opposite shore. Slowly the alligator's heart rate began to recover. In ten minutes it increased to 5 beats per minute. Within 30 minutes it returned to the previous value. What could this mean? Was this normal? It was obvious that Will frightened the alligator. His canoe passed within a few yards of the alligator, yet its heart slowed instead of speeding up. Everyone knows heart rate speeds up when frightened. What is going on? What am I missing?

I remained in the blind the rest of the day and recorded data. Everything checked out, but the anomalous data haunted me. I replayed the event in my mind a hundred times. I had confidence in the telemetry measurements. The zoo study had affirmed its reliability. The alligator was healthy. Just the night before it had caught a raccoon. I knew enough about science to know the data that doesn't fit preconceived notions is often the most exciting. I also understood the importance of accidents in research. Fleming discovered penicillin by accident. The discovery of X-rays was accidental. Yet accidents alone mean little. A successful scientist must understand the significance of the accidental observation and have the imagination and tenacity to follow through. It takes insight and considerable effort to convert an accidental observation into a significant discovery. It is unlikely that Dr. Fleming was the first microbiologist to have a bacterial culture contaminated with a common mold. He discovered penicillin because he had the insight to wonder why there was a ring around the mold where no bacteria could grow. I had made an accidental observation, but I needed more research experience. Here was an extraordinarily important piece to a puzzle, but I was uncertain where the piece fit, or even to what puzzle it belonged.

Two events were inexorably united: the alligator was frightened by the approach of the canoe and it slowed its heart. My budding research philosophy was simple: animal responses are functional. Just because we do not understand a particular structure or function or behavior means nothing. The structure or function or behavior is important to its owner or it wouldn't have it. There once was a list of 186 human organs thought to be useless relics of our evolutionary past. As knowledge replaced ignorance, that list dwindled to oblivion. All our parts function. The alligator had a functional reason for slowing its heart. The response was adaptive. My task was to discover how it was useful. It makes good physiological sense for a submerged alligator to slow its heart. It is safe underwater and slowing its heart prolongs the time it can spend submerged. Could this accidental observation be somehow related to the classic diving response?

Then it happened. I was instantly overwhelmed. Strong emotions flooded over me like breakers on a sandcastle. Excitement, awe, astonishment, and humility overpowered me one at a time and simultaneously. It was reminiscent to a person's life flashing before them at the moment of impending doom. Because of this accidental discovery, I understood a response others had overlooked. It was too obvious. The alligator slowed its heart in response to fear, not submergence. My measurements made while the alligator was submerged showed a normal heart rate because it was not frightened. There had been two exceptions: this one with Will and the canoe, and when I first released the alligator. During both events the alligator was frightened and its heart slowed. I had wrongly attributed the response upon release to diving. I had learned my textbook physiology well, but the textbooks were wrong!

Other workers studying alligator diving in the laboratory made the same wrong conclusions. The alligators' hearts slowed when they were forcibly submerged because they were frightened. The response had nothing to do with diving. My telemetry measurements of diving with a free-ranging alligator clearly showed that submergence alone had no effect on heart rate.

The entire diving bradycardia literature needed to be re-evaluated. Is it possible that a lowly graduate student could see something missed by the giants of science? Fear soon replaced excitement. Who would believe me? I must have additional documentation. More experimental evidence was necessary. During another sleepless night my mind continued to race. What experiments could I perform? I had it. I must repeat the accidental discovery under carefully controlled conditions. I was onto something, but I needed proof. A single episode means little; it could be a fluke. Graduate students often jokingly refer to research as re-search because experiments must be repeated again and again.

The next day three related experiments were performed. The first was not unlike yesterday's accident. I monitored the alligator's heart rate and directed my assistant to pole the boat quietly to within about 20 yards of the submerged alligator. On his command he pounded on the bottom of the aluminum boat with an oar. As my new theory predicted, there was an immediate and profound drop in the alligator's heart rate. It was fear and not submergence that triggered the slowing of the alligator's heart. Perhaps the response was peculiar to a threat from the water. I wanted to see if vibrations from the bank elicited the same response. My assistant monitored the alligator and I popped wheelies on my 350 cc Honda motorcycle. Certainly, this would vibrate the ground. Again, the disturbance triggered an immediate slowing of the alligator's heart. The last experiment was unplanned, but instructive. Late in the afternoon the alligator's heart suddenly slowed again. Soon a graduate student drove up to do a bit of sport fishing. The alligator had detected vibration (or sound) from the car before I did! Confirmation was sweet.

Science makes progress as new theories are tested and either confirmed or rejected. A new theory, no matter how profound or glamorous, is of no value unless predictions made from the theory prove valid. My new theory predicted that alligators respond to fear by slowing their hearts. The new data supported the new theory. So far, so good. Where was this new theory going to lead? What did it really mean in the grand scheme of things? Did it have any important applications to other animals, or to man?

Over the next several days additional evidence was gathered confirming a slowing of the alligator's heart in response by fear. By summer's end, I had collected data from over 200 evoked fear episodes with four different alligators. Each succeeding experiment with alligators in the water added strength to my theory. Data from caged alligators forced a refinement of the theory. Caged alligators responded to human approach with aggressive behavior and increased heart rate. The alligator must be frightened and underwater to exhibit the response. It must have a safe hiding place. My theory had gained enough credibility to be published.

It was a good summer and I learned more than I had ever dreamed possible. My most important discovery was made by accident, but I had successfully followed through on the observation. As often happens in science, my accidental discovery had nothing to do with the thermoregulation I had set out to explore. The discovery about the relation between fear and heart rate dwarfed the thermoregulation findings and charted the future direction of my research. A myriad of questions demanded answers. How widespread was this seemingly anomalous cardiovascular response to fear? How did it help the animal survive? Scientific research always raises more questions than it answers. That is one of its major attractions.

# 4

# Retreating woodchuck

*Stay, stay at home, my heart, and rest;*
*Home-keeping hearts are happiest,*
*For those that wander they know not where*
*Are full of trouble and full of care;*
*To stay at home is best.*

—Henry Wadsworth Longfellow

Five years have passed since the accidental discovery at Big Lake. I completed my Master's Degree at Baylor University, studied two quarters at UCLA, and completed my doctorate in Zoology from Texas Tech University. After teaching one summer at Fort Hays Kansas State University and a year at Rochester Institute of Technology in up-state New York, I took a teaching/research position at Northeastern Oklahoma State University at Tahlequah.

Eastern Oklahoma is as different from western Oklahoma as rainforest is from desert. It rains nearly 40 inches per year in eastern Oklahoma compared to less than 20 inches on the grassy plains of western Oklahoma. Babbling brooks of clear water, scenic rivers and picturesque lakes are in a perpetual bidding war to attract visitors. The rolling hills of the eastern half of the state are covered with a mosaic of mixed hardwood trees. Northern bird species spend mild winters here. Wildlife photographers and bird watchers are attracted to wintering bald eagles and other birds of prey. White pine, red cedar, evergreen fern and native holly add green to the coldest days. An explosion of color punctuates the end of winter as native red bud and dogwood trees compete for brilliance. Tourists from around the world arrive by bus for annual "Dogwood Tours."

Water sports also attract thousands of visitors. Clear streams and rushing rivers allure canoeists and rafters. Fishing is excellent throughout the year. Summers are filled with water skiing, bass tournaments and an unending profusion of beautiful wild flowers. The end of summer is marked by a second color explosion and by the return of tourist buses for the ever-popular "Fall Foliage Tours." Splashes of yellow from sycamore, vivid reds of sumac and earthy browns from oak reward the weary traveler. Eastern Oklahoma is truly a field zoologist's paradise.

The Illinois River is a nationally recognized Scenic River, and Tenkiller is among the most beautiful lakes in North America, surrounded by white pine and red cedar-covered limestone outcroppings. The water is deep and clear enough for scuba diving. Three days after moving to Tahlequah, we bought a ski boat. It would be sacrilegious not to be on such a magnificent lake each weekend. That ski boat eventually provided transportation to and from research study sites. Serendipity was still alive and well.

I had been active in research, but my studies had been directed toward alligator thermoregulation, mostly in the library. The alligator growth study continued without interruption, and many hours had been spent going over the deviant fear response data. The alligator heart rate research was done backwards. Normally one researches the scientific literature BEFORE the study. With my accidental discovery this was impossible. I approached the literature after already collecting conclusive data that fear evoked a decrease in heart rate of wild alligators. Much to my surprise, scattered throughout the scientific literature were dozens of isolated accounts of animals "freezing" and slowing their hearts in response to sudden noise, movement or other threatening stimuli. Field mice slow their heart rate and even lower their body temperature when a shadow passes overhead. Ground squirrels slow their hearts while hiding. Newborn deer fawns drop to the ground, freeze, and slow their hearts when approached by man or dog. Some captive fish stopped their hearts during a minor California earthquake. Supporting data was in the literature. It was just scattered. Perhaps I could be the first to study this response in detail. Maybe I had found my niche.

My discovery at Welder, along with the early work of Dr. Geir Gabrielsen of Norway, helped draw attention to the importance of using radio telemetry to study free ranging animals under natural conditions. Wild animals tested in laboratories respond abnormally, and many of the early diving studies needed to be re-evaluated in the light of recent telemetry studies. Even seals and diving ducks were found to maintain a normal heart rate during short voluntary feeding dives.

Investigators studying diving animals in the laboratory had drawn the wrong conclusion. They unknowingly recorded a response to fear and not a response to submergence. Even in Dr. Scholander's classic paper about diving bradycardia, he mentioned that sometimes seals responded to noise with a marked reduction of heart rate. Surprisingly, no one had set out to investigate this response. All the studies reported the response in passing while investigating something else. In retrospect, it seems obvious that forcing a wild animal underwater in a laboratory would frighten it.

My preliminary data with alligators suggested the entire "diving bradycardia" response might be a modification of a broader passive defense response to fear. During short voluntary dives the animals are not frightened nor is there a need to conserve oxygen, thus heart rate remains at pre-dive values. What about non-diving animals? I needed a new model to study.

A model has many definitions in science. To a chemist it may be three-dimensional representations of molecules consisting of colored wooden or plastic spheres and connecting rods. For a physicist a model may be a mathematical formula describing details of some physical phenomenon. A physiologist sees a model as an animal that exhibits the response to be studied. For example, much of what we know about human kidney function was discovered using the kangaroo rat as a model. Kangaroo rats are excellent models for kidney research because they can produce urine three times more salty then seawater.

Alligators had been good to me. They had helped me earn two graduate degrees and publish over a dozen research papers, but I wanted to see if the passive fear response occurred in other animals. I wanted to study some plentiful animal that would retreat to a nest or burrow when frightened. Most reptiles and mice are too small for the kind of radio telemetry I needed to use. Birds fly away when frightened. Rabbits and squirrels only use their nests to rear their young. Gophers and moles are common, but spend little time above ground and thus are difficult to observe. In western Oklahoma prairie dogs would be a good choice, but they are not plentiful in eastern Oklahoma.

After careful consideration of the abundant local animals, I decided to continue studying the passive fear response using woodchucks. Woodchucks are abundant and easy to catch. They have limited home ranges near their borrows, so they would stay in radio range. They are interesting creatures and their alleged weather predicting ability has influenced American folklore by giving us Groundhog Day (February 2nd). They inhabit wooded areas throughout most of the eastern United States and southern Canada. Woodchucks are often seen along roadsides and in clearings. They dig extensive underground burrows and are expert hibernators. In Canada, where winters are long and cold, they spend 75% of their life in hibernation. Most importantly they spend a lot of time above ground foraging for food, but retreat to the safety of their burrow when frightened.

The selection of the study site was easy. The ideal study site must provide natural habitat for the animal to be studied, must be accessible, yet protected from hunting and trapping. I had learned at Welder that disturbed animals sometimes leave their home territory. Although I had access to both private and public land near Tahlequah, I decided to use a small island in Tenkiller Lake. Buzzard Island, as locals called it, was accessible only by boat and provided excellent woodchuck habitat.

A simpler telemetry transmitter was needed—one that would monitor the electrocardiogram, respiration and body temperature. Advances in integrated circuits and my past telemetry experience made the new design easier. Within weeks I had a new system that included a memory circuit that was switched on and off by placing a magnet near the animal, helping batteries last several months. The finished transmitter was about the diameter of a dime and a little over an inch in length. Other telemetry systems on the market at the time cost tens of thousands of dollars because they required special broadband receivers and multichannel tape recorders for recording ECG. Mine cost less than twenty dollars and used an inexpensive FM radio and cassette recorder. The system worked well and often provided a range exceeding 100 yards. It was published, and today several people are using modified versions of the system I designed. All I needed were some volunteer woodchucks.

While awaiting major National Science Foundation funding, I applied for and was provisionally awarded a small pilot grant from the university. The university Dean, who controlled pilot grants, liked woodchucks. She had some living near her garden and enjoyed watching them. Two she had even named. She considered them friends and was reluctant to fund my study if the woodchucks would be hurt or killed. I assured her they would be well cared for and detained in the laboratory no longer than necessary. With her fears laid to rest, I was awarded a few hundred dollars for the purchase of holding pens and traps. The required collecting permits were obtained, traps purchased and a new chapter of research began.

Teaching loads are heavy at small state universities, and without the help of scores of eager students my research would have died on the vine. My students were enthusiastic helpers. What better way to learn about science than by actually doing science? From the outset excellent students surrounded me. Some were pre-med students and are now practicing medicine throughout the country. Others were biology majors and have completed graduate degrees and are now teaching at major universities. A few were education majors and are currently teaching

Biology throughout Oklahoma and the surrounding states. They certainly have some entertaining adventures to inspire another generation of students.

Dr. Fred White of UCLA had taught me the importance of observing study animals in their natural environment before attempting any research. Several students and I located nearby areas where woodchucks thrived. We constructed makeshift blinds and began observing woodchuck behavior. Woodchucks dig extensive burrows and often live near other woodchucks, although social interaction is rare. They seldom wander far from their burrows, and make a rapid retreat when threatened. They are most active early in the morning and late in the afternoon. They apparently sleep inside the burrows at night and during mid-day. After two weeks of careful observation, we began trapping.

As soon as we had woodchucks in the laboratory, the research Dean sent an Emissary to check on the captives. One woodchuck was particularly plump and relatively easy to handle. I had often picked it up by the back of the neck to move it from one cage to another while cleaning cages. When the Emissary arrived she wanted to see if the "cute little woodchucks" had food and water. I showed her their comfortable living quarters with ample food and fresh water. I tried to impress her with their tameness. Money and research were at stake.

Attempting to make a good impression to assure funding can, at times, be inconsistent with good judgment. Such was the case here. I decided it might appear cowardly or cruel to grab the woodchuck by the back of the neck. And I needed the funding! I picked the woodchuck up as one would pick up a pet cat and cradled it in my arm. It immediately located the arm that held it and bit down—hard! I was wearing a lab coat with the sleeve rolled about two thirds of the way up my arm. Luckily for my arm, the woodchuck was biting through six thicknesses of lab coat. Nevertheless, woodchucks have strong gnawing teeth and bite very hard. And this woodchuck was determined to inflict bodily harm. The Emissary remarked that well cared-for captives do not bite the hand that feeds them. I brushed her concern aside and convinced her it was merely playing: akin to the mock biting of a puppy. Perspiration was pouring from my forehead due to the intense pain. I felt nauseous, but managed a weak smile as I walked the Emissary to the door and extended my regards to the generous Dean. The Emissary was impressed with my woodchuck husbandry and would give a good report. The pain was excruciating.

With the Emissary out of earshot, I instantly returned the animal to its cage and pried it off my arm. I sat down from weakness and nearly passed out. I swore softly. The woodchuck had inflicted a deep bruise about two inches in diameter and I was bleeding, but I had saved the grant. Research requires dedication and self-discipline. I had both. I also had a very sore arm for several days.

Soon transmitters were attached and the animals were released on Buzzard Island at Tenkiller Lake. Buzzard Island is a 20-acre hill surrounded by water and provides excellent woodchuck habitat. The island slopes gradually from the north and west and rises sharply out of the lake on the south and east. The southeast corner is capped with an outcropping of limestone rocks and is about 50 feet above the lake. The name of the island is fitting, as several turkey vultures, called buzzards in this region, roosted each night in a large dead oak tree near the top of the limestone outcropping. There was evidence of a long abandoned nest site. Daily boating activity had forced the vultures to find a more secluded place to fledge their young.

An array of antennas was set up on the island near a large outcropping of rocks. Cables connected them to a switch box and radio. A tree house blind in view of the rocks proved useful. Walkie-talkies were used to communicate between blind and boat. Most of the measurements were taken from a large pontoon boat purchased for research. My ski boat was used to shuttle students and equipment to and from the island.

Experiment day finally arrived. Several woodchuck experiments were planned. Four telemetered woodchucks were on the island and all had burrows near the rocks. Woody Woodruff, a pre-med student, spent the night. Since the woodchucks slept in their burrows, measurements were made at 2-hour intervals so Woody could get some needed sleep. I wanted to arrive before daybreak to check everything and go over the proposed experiments with Woody. Preliminary observations had indicated a 60 percent reduction in heart rate each time the woodchuck retreated to its burrow. We were going to attempt several experiments with each of the four animals. Lady, my Border Collie, was invited. She enjoyed research and had helped catch several deer fawns for a study at Welder. She never bit them, but would hold them by throwing her body on them.

There was little evidence of the new day when I arrived at the lake before 5:00 AM. Radio contact confirmed all was well at the island. I loaded some additional equipment onto the boat and we departed. Lady and I glided across the glassy surface to the pontoon boat. The eastern sky was graying as I stowed the last of the equipment. The flickering of gas lanterns revealed the presence of fishing boats as far as the eye could see. Some fishermen were putting their boats in the water while others were leaving. The dawn was quiet. A distant boat broke the silence. Muted voices could be heard over the water and a chorus of spring peeper tree frogs called from the woods. The din of crickets and katydids blended in the morning mist. The wakes of passing boats slowly reached the beach and splashed sleepily against the rocks.

As the sun appeared the sound of distant frogs and crickets gave way to the piercing bickering of blue jays and the soft cooing of mourning doves. A robin and common flicker were foraging on the ground. A solitary swamp rabbit slowly hopped in a clearing where lightning killed a mighty oak tree the previous year. A tight formation of four crows flew overhead. The call betrayed their identity as fish crows. Although smaller than common crows, fish crows are most easily identified by their nasal call. The woodchucks were still inside their burrows. Squirrels began scampering along the ground in search of acorns. It wouldn't be long.

The first indication of woodchuck activity was increased heart rate from the transmitter. Soon the signal alternately grew stronger and weaker, indicating movement. As the animal came above ground the signal grew louder. Woody radioed that the animal was basking in the morning sun just outside its burrow. Within 15 minutes all four animals were enjoying the sun. Shortly, two of the animals left the rocks to forage for food. That was our signal to begin work.

Several minutes of undisturbed heart rate data was recorded as a control for the experiment. Woody was instructed to approach the animal. When he was about 20 yards from the woodchuck, it bolted toward its burrow. During the race home its heart rate increased, but as soon as it retreated inside its burrow the heart rate dropped to half the undisturbed value. Comparable experiments were performed on the other woodchucks with identical results. If the animal was disturbed away from its den, heart rate increased, but as soon as it retreated to safety, a marked slowing of its heart was observed. Alligators showed a similar response. Caged alligators increased their heart rates when disturbed, but if they were in the safety of the water their hearts slowed when frightened.

Each animal was frightened twice before we retreated to the pontoon boat. We ate a leisurely lunch, fished half-heartedly and discussed our results. Late in the afternoon, when the woodchucks were again out and about, we started a second series of experiments. This time the dog was used to frighten the woodchuck. Lady soon picked up the scent and gave chase. Again the woodchuck's heart raced as it scurried to the burrow. Upon entering its burrow, the heart rate plummeted to about half the pre-disturbance values as before. Lady continued to search for the quarry and after about five minutes found the entrance to the woodchuck's burrow. As soon as she began digging, the woodchuck's heart rate dropped even more and remained very slow until the dog was called back to the boat. Within seven minutes the heart rate returned to the previous undisturbed value. Soon the woodchuck was again foraging.

These data were particularly exciting. Here was another piece of the puzzle. Until the last experiment with the dog, I had thought the passive fear response had an all-or-nothing effect on heart rate. That is, I had assumed perceived fear by the animal induced maximal heart slowing. I was wrong. The amount of slowing is proportional to the amount of stimulus. The stronger the stimulus the slower the heart rate. If I had not used the dog and if she had not found the burrow and begun digging, this important piece of the puzzle would have escaped notice. Luck continued to smile on my research. Many more woodchuck experiments were performed during the summer and each experiment added credibility to the existing data.

What a pleasant way to spend a summer. The results were given by my students at an international telemetry conference at Oxford University in England. The fear data was well received and I was able to meet some of the world leaders in radio telemetry. Several enduring friendships were made.

Although I still did not know how widespread the passive fear response was, I had excellent data for two species under natural conditions in their native habitat. Yet an important question remained. How did this response help the woodchuck survive? For the alligator, holding its breath underwater slows its metabolism, which increases its dive time and thus enhances survival. What about the woodchuck? A quick review of the scientific literature for gophers showed that oxygen levels can drop from the normal 21% in air to as low as 5% inside plugged burrows. Many burrowing animals plug their burrows when attacked. A woodchuck inside its burrow is safe, but may face the same problem as a submerged alligator. The supply of oxygen available in the burrow may be scarce. Could this be but a modification of the dive response to conserve oxygen? What about a non-diving, non-burrowing animal? More studies with more species were needed.

As I continued collecting pieces to the puzzle, I wished I knew what kind of picture the pieces were going to make. Not knowing was keeping me awake nights. It was becoming an obsession…and I loved it. Even higher peaks were beckoning me ever upward.

# 5

# Tip of the iceberg

*Let your imagination go, guiding it by judgment and principle, but holding it in and directing it by experiment. Nature is your best friend and critic in experimental science if you only allow her intimations to fall unbiased on your mind. Nothing is so good as an experiment which, while it sets an error right, gives you as a reward for your humility an absolute advance in knowledge.*

—Michael Faraday

Finding a rabbit using radio telemetry

In the hindsight of many years, neither alligator nor woodchuck yielded the most pieces to the passive fear puzzle. Instead, rabbits provided more answers than all others combined. I explored the passive fear response in eastern cotton-tails, domestic rabbits and swamp rabbits for five years. Two observations prompted the use of rabbits. I remembered that, when I was a young boy, my uncle and grandfather sometimes used a curious technique to hunt jackrabbits at night. Occasionally the rabbits became a nuisance and competed with cattle for pasture. They were so abundant that over a hundred could be killed in a single night on our 80-acre pasture. My kin would blind the rabbits with a spotlight and discharge a shotgun over the rabbit's heads. The rabbits were "paralyzed with fear" and could be picked up by hand. None were wasted, and to this day I still enjoy a good rabbit stew. Ranchers in west Texas also used this unusual hunting technique, with one variation. They used fishnets to grab the frightened rabbits.

The second vote for rabbits came from observations made during the wood-chuck study. Swamp rabbits inhabited Buzzard Island and were observed along with woodchuck. When frightened, the woodchucks retreated to the safety of their burrows. Rabbits exposed to the same stimulus would often simply lower their ears, crouch down, and remain motionless. This was an obvious hiding response and I wanted to monitor their heart rate during such a simple hiding episode. Hiding from danger was a common thread tying many species together. Alligators only showed the response if they had access to water in which to hide; woodchucks slowed their hearts only when they retreated inside their burrows. What about rabbits?

Swamp rabbit

Swamp rabbits are abundant in eastern Oklahoma, but they proved difficult to trap. We captured eastern cottontail rabbits with ease and decided to begin the next phase of the study with them. Most everyone is familiar with cottontails; they are abundant, lovable creatures found throughout much of the United States. Like woodchuck, they have earned an endearing place in western folklore. Peter Cottontail and Bugs Bunny have entertained generations of children. And then there is the ever-elusive Easter Bunny. Rabbits are wonderful characters for children's stories and cartoons.

Although some significant discoveries were made with cottontails, they were not good experimental animals. In spite of their docile image, cottontails were difficult to maintain in captivity and often remained incorrigible. Some did not eat in captivity and were released to avoid starvation. Others became so frightened when we entered the room they injured themselves trying to escape. Nevertheless, we were able to telemeter 12 cottontails, release them on Buzzard Island at Tenkiller, and make some astonishing discoveries.

Unlike woodchucks, rabbits did not retreat to the safety of a burrow. If cottontail rabbits were frightened while in the open, they fled and their hearts raced. This is to be expected from any fleeing animal, and was identical to the response of caged alligators, and woodchucks away from their burrows. Basic physiology demands that it be this way. Greater exertion results in greater need for oxygen and blood flow. The heart rate must increase to meet these greater demands. If, instead, the rabbits were frightened and had cover, they would immediately lower their ears and crouch low to the ground. They were hiding *in situ*. Heart rate slowed the moment they lowered their profile. If the threatening investigator or dog backed off, the heart rate returned to pre-stimulus values within five minutes or less, and pre-disturbance activity was resumed.

Even fleeing rabbits, like frightened woodchucks, often ran for cover. Rabbits would seek seclusion in a depression in the ground or behind a log or inside a briar. Upon reaching cover they would hide by assuming the low profile, resulting once again in a lowered heart rate. If the investigator or dog approached to within a few feet of the rabbit, it would suddenly bolt to safety. Heart rate could double within a single heart beat interval. The rapid acceleration of heart rate raised some important questions regarding the control of heart rate in rabbits. Some of the questions demanded answers that could only be obtained in the laboratory. Field work was placed on temporary hold.

To understand the next series of experiments, we must first consider how the heart beats and how the brain controls it. One of the striking features of cardiac muscle cells is their ability to beat without an external stimulus. Each individual cell has its own cadence and will contract in isolation from other heart cells. The same is true for the entire heart. If the heart is removed from the body and placed in an oxygenated nutrient solution, it will continue to beat. Skeletal muscles, in contrast, require a nervous impulse from the brain for each contraction. The pacemaker region of the heart sets the heart rhythm because it has the fastest cadence. The rest of the heart falls in step with the pacemaker.

Although the heart beats without an impulse from the brain, heart rate is controlled by the central nervous system. There are two neural connections between brain and heart. These pathways are part of the autonomic nervous system. The two subdivisions are called the sympathetic and parasympathetic portions of the autonomic nervous system. They have long been known to have far reaching, but opposite, effects on most internal organs of the body. They perform basic control functions and work without conscious control. The autonomic nervous system helps digest food, regulates blood pressure, controls body temperature and a host of other vital functions. Specifically, there are two ways the autonomic nervous system affects heart rate. The sympathetic portion of the autonomic nervous system releases adrenaline near the pacemaker of the heart, accelerating heart rate. In contrast, the parasympathetic pathway releases acetylcholine, which slows the heart.

Conventional wisdom suggested rabbits slowed their hearts by increasing the release of acetylcholine, and increased heart rate by releasing more adrenaline near the pacemaker. I wanted to determine the relative importance of the two portions of the autonomic nervous system in the passive fear response. One could surgically abolish either portion of the system, but surgery was permanent and I wanted to release the experimental animals after the study. Neural blocking agents were available and safe to use. The effects of the neural blockers wore off, and the animals could be reused in other experiments or released. Propranolol was used to block the release of sympathetic adrenaline, and atropine was given to block the release of parasympathetic acetylcholine.

Thus far, the most obvious physiological indicator of the passive fear response was the marked reduction of heart rate when the rabbit crouched and hid. Logically, the parasympathetic portion of the autonomic nervous system was slowing the heart. After determining the proper dose and drug duration, a telemetered cottontail was injected with atropine. As expected, the resting heart rate increased with the removal of the braking action of acetylcholine. Rabbit behavior appeared normal and the caged animal was frightened by the approach of the investigator. Immediately upon crouching, the heart rate dropped! This was unexpected, but I had learned long ago to welcome the unexpected.

A careful examination of the literature revealed that both portions of the autonomic nervous system are normally active all the time. There is a simultaneous acceleration by adrenaline release and a deceleration effect with acetylcholine release. Under normal conditions small quantities of both substances are continually released. The relative amounts of the neural transmitters released are varied to control heart rate. This means either portion of the autonomic nervous system could both increase and decrease heart rate. The heart could be accelerated by increasing the amount of adrenaline released or by reducing acetylcholine release. Conversely, reduced adrenaline release or increased acetylcholine release could slow the heart. Complex, yet beautifully simple.

An analogy is useful. Picture someone driving a car on a level road with one foot pressed lightly on the accelerator and the other foot pressing slightly on the brake (This procedure is only recommended by unscrupulous brake repair shops). Speed may be increased either by pressing harder on the gas or letting up on the brake. Conversely, the car may be slowed by letting up on the gas or by increasing brake pressure. Either foot could speed up or slow down the car. So it is with the two portions of the autonomic nervous system; either portion can increase or decrease heart rate.

This accounts for the rabbit's ability to slow its heart when atropine blocked the release of acetylcholine. A careful examination of the telemetered ECG revealed a second atropine effect. Although the stimulus slowed the heart in the atropine-treated rabbit, the rate of deceleration was reduced. The heart could still be slowed the same amount but it required nearly 10 seconds with atropine treatment, compared to less than 1 second in the untreated control rabbits.

Next we tested rabbits injected with propranolol to block the sympathetic portion of the autonomic nervous system. As expected, removal of the normal accelerator depressed resting heart rate. Behavior again appeared normal. Once again a stimulus resulted in the crouch-and-hide response. As expected, the heart again slowed. This information plus that gained by atropine treatment provided new and significant pieces of the passive fear puzzle. Either portion of the autonomic nervous system alone could reduce the heart rate of frightened rabbits, but both systems together were necessary for the rapid changes observed in untreated, free ranging rabbits.

As a final check, rabbits in the laboratory were injected with both propranolol and atropine. Behavior appeared normal, including the crouch-and-hide response, but heart rate remained unchanged. This was reassuring, because when both drugs are used together, the heart and brain are effectively disconnected. No known mechanism could alter heart rate under these circumstances.

Unfortunately, many investigators bring wild animals into the laboratory and simply assume they behave and respond as they would in the wild. This can be misleading, as was the case for other scientists for over 50 years with forced diving experiments in the laboratory. It was the effect of fear, not submergence, they were seeing. If I had not first studied free ranging cottontails, I would have had little confidence in the laboratory results. I had already learned the heart rate response of wild rabbits to threat was a marked slowing of their hearts and was able to duplicate those results in the laboratory. The field study added credibility to the laboratory study. And the laboratory work added valuable information unavailable in the field. Since the results of untreated rabbits were similar in the field and lab, drug treatment experiments in the lab were defensible. Broad features of the passive fear response picture were slowly emerging from the puzzle pieces and no one on earth had seen this picture before.

Why study wild animals? As mentioned, wild cottontail rabbits were difficult to trap and sometimes impossible to maintain in the laboratory. Why not use tame rabbits? This seemed logical, so several large, tame rabbits were secured and transmitters attached. There was an insurmountable problem. Tame rabbits could not be reliably frightened. Generations of selection had resulted in rabbits that no longer could be frightened. Due to selective breeding for docility, they had lost their fear response that is so important for wild animals. Neither man, nor dog, nor the sound of a gun or firecracker resulted in either the typical crouch-and-hide behavior or a change in heart rate.

Nevertheless, this side excursion was educational. It is an unfortunate fact of medical research that nearly all behavioral and physiological studies are completed using "laboratory animals." Common laboratory animals include white mice, white rats, guinea pigs and tame rabbits. These animals have been used in medical and psychological research for decades. An important unnatural selection has occurred. The breeders of laboratory animals select for docility. Often their animals are advertised as "non-aggressive and easy to handle." Fearful or aggressive animals are eliminated. After countless generations of selective breeding, the result is that most laboratory animals have lost their fear response. They are still useful for some kinds of research, but I could not use them for my study of passive fear. This does raise some important questions regarding behavioral studies, especially if conclusions are to be applied to wild species. In a sense, laboratory animals are fake animals. True, they live and breathe and reproduce, but their survival instincts have been lost by generations of unnatural selection for ease of handling.

My next candidate in the search to better understand the passive defense response was swamp rabbits. Although they are common in eastern Oklahoma, they proved difficult to trap. I called an old friend, Ted Joanen, Research Director of the Rockefeller Wildlife Refuge in coastal Louisiana. His reply was simply, "How many do you want and when do you need them?" A quick trip to Louisiana yielded 22 large healthy specimens. Networking is as important in zoology as it is in the business world.

Unlike cottontails, swamp rabbits proved to be excellent study animals. They were large and rugged and readily adapted to laboratory life. Swamp rabbits were used in a variety of ways. They were used to confirm the neural blocking experiments started with cottontails. The results were similar, but larger numbers of animals increased credibility. Some new studies were undertaken. Several telemetered swamp rabbits were released on the island, and field observations supported the earlier work with cottontails. Again larger numbers or rabbits and more replicate experiments enhanced the reliability of my results.

In the grand scheme of things, rabbits were important because they helped me begin to see the passive fear response in a new light. Rabbits, like alligators and woodchucks, slowed their hearts when frightened, but they did not lack oxygen. They were breathing normal air, yet still slowed their hearts when frightened and hiding. Perhaps the passive fear response was even more widespread than I had thought. Could the classic diving response possibly be nothing more than a specialized and modified passive fear response? This theory would provide additional food for thought, but it had to take a number and wait in line. Due to the similarity of the passive fear response to the classic diving response, I decided to look at oxygen consumption during passive fear next.

Oxygen consumption is a measure of total metabolic activity and normally requires collecting all expired air. Humans simply breathe into a tube and the expired air is analyzed. I was able to fasten a mask to alligators to measure metabolism, but wild rabbits would not breathe into a tube or tolerate a mask. Steady state oxygen consumption could be determined without a mask by slowly passing air through a small, airtight cage. Oxygen consumption was then calculated by measuring the rate of flow and the different concentration of oxygen in the air entering and leaving the cage. I wanted to monitor a sudden change in oxygen consumption. The best I could do was force the rabbit inside a small chamber made from large plastic pipe. After several practice trials, the rabbits acclimated well enough to be tested. Changes in oxygen consumption could be measured within 30 seconds of the stimulus. Heart rate dropped as it did in the field, again adding credibility to this highly artificial experiment. During a five-minute crouch-and-hide period, oxygen consumption decreased 60 percent. Another piece of the puzzle was found. And perhaps this was the most important piece yet found, for it suggested far-reaching physiological effects of the passive fear response.

It was apparent that heart rate reduction was but the tip of the iceberg. Respiration and oxygen consumption were also reduced. The passive fear response entailed a suite of highly orchestrated physiological responses. It was perhaps more complex and at least as widespread as the active fear response. It was observed in both predator and prey species.

The question of its adaptive significance was always on my mind. Why did different animals living in different environments show the same response? Could it be important in humans? The heart rate response of humans to a sudden noise is variable. Perhaps this is due to a different emotional response. Anger might increase heart rate. Perhaps fear, under the right circumstances, could account for a decrease in heart rate. The adaptive significance was perhaps the most important piece of the puzzle. The question haunted me day and night. More species had to be studied. Or, was I overlooking the obvious? To how many different puzzles did I hold pieces?

# 6

# Diving rabbits

*It is not the horse that draws the cart, but the oats.*

—Russian proverb

It appeared that I remained the only scientist in the world investigating, by design, the passive fear response of wild animals. That was both good and bad. It was nice not having to look over my shoulder for fear someone else would discover some aspect of the passive fear response and publish it before I did. Yet it would have been nice to have confirmations of my results by others, or simply to have had another scientist's viewpoint. No doubt I would soon have company, for the response seemed widespread. Working in my favor, it seemed few field zoologists had the necessary electronics background to design their own telemetry systems.

While attempting to document how widespread passive fear was among various animal species and its implications for survival, pressing questions about diving physiology demanded answers. My animal model might provide additional insight into both puzzles, or maybe there was but one big picture. I had a strong hunch passive fear and diving were somehow related, but did I have sufficient insight and imagination to figure out how?

My accidental discovery with alligators that fear, NOT submergence, caused their heart rate to slow added new insight and seemed to show a relation between passive fear and forced diving. It also raised serious doubts about hundreds of previously published studies about the validity of forced diving experiments of wild animals under unnatural laboratory conditions. Dr. Geir Gabrielsen of Norway and others were starting to re-think much of the older research.

In retrospect, it seems odd how anyone would think the response of a wild animal tied to a tilting board and forcibly submerged underwater would provide useful information about the physiology of natural diving. Such harsh treatment would evoke a fear of drowning. Many of the older studies needed to be repeated under natural free-ranging conditions, and modern advances in physiological telemetry made such studies straightforward. Yet there is a place for laboratory experiments. What was needed was a comparison of forced diving and natural diving with another animal model. My discoveries with alligators needed additional support with another species to show the response was widespread. Only then could some of these burning questions be answered. Perhaps I already had the perfect animal model in my laboratory.

Swamp rabbits are excellent swimmers and good divers. Once, while collecting alligators in East Texas, I encountered a swamp rabbit swimming in the creek ahead of my boat. When it saw the boat, it submerged and swam to the opposite bank. It covered at least ten yards underwater! That appeared an amazing stunt to this western Oklahoma farm boy. As their name suggests, swamp rabbits are well suited for life around water. I already knew more about their passive fear response than any other animal. Perhaps an investigation of their physiological response to forced and voluntary diving would be instructive.

As it turned out, swamp rabbits were good models for studying diving. I wanted to begin by duplicating the old published forced diving experiments that had been done with many diving animals. Nothing had been published about diving in swamp rabbits. In research, as in life, it is best to travel from what is known into the unknown by small steps. Although I felt strongly that forced diving under artificial laboratory conditions implied little about voluntary diving in free-ranging animals, it was where I had to begin. My ultimate goal was to better understand diving under natural conditions, but first I needed to repeat the work of a host of other investigators using my new model. This was necessary for comparison and as a starting point. In a sense, this was to be my control experiment.

The experimental procedure was simple. A telemetered swamp rabbit was placed in a small wire cage and forcibly submerged underwater for ten seconds while its heart rate was monitored. Submergence brought an immediate and profound drop in heart rate, in spite of continuous struggling by the rabbit. These results were like those reported in the literature for dozens of forcibly submerged species. I expected as much, but still needed the experimental support. The experiment was repeated again and again using other rabbits. Each experiment yielded similar results. Since I had used neural blocking agents to study the passive fear response, it was logical to use them during forced diving experiments as well.

Propranolol was used to block the release of sympathetic adrenaline and, as expected, had no significant effect. Heart rate still slowed during forced diving. Next, atropine was used to block the release of parasympathetic acetylcholine. Here the results were surprising. In the passive fear response of cottontails and swamp rabbits, neither propranolol nor atropine alone could block the slowing of the heart. With forced diving, atropine blocked the response! Here was a new piece, but I was not certain to which puzzle it belonged. It appeared the diving response was mediated solely by the parasympathetic portion of the autonomic nervous system. Careful re-evaluation of all diving data revealed two other differences between diving and passive fear. The rate at which the diving rabbit's heart slowed was more gradual with diving, yet the degree of slowing was greater. In other words, during forced diving, the rabbit's heart rate decreased more slowly, but ended up at a lower heart rate, than when free ranging rabbits were frightened by the approach of man or dog. And, surprisingly, this occurred in spite of vigorous attempts by the rabbit to escape during the dive. Somehow heart rate seemed not to be responding to oxygen demands by the very active rabbit. These differences were statistically significant and were trying to tell me something important. It was reasonable to conclude that fear was involved in both responses. In passive fear, it was fear of a possible predator. In forced diving, there was the obvious fear of drowning. In the former, all outward activity ceased; the rabbit remained motionless. During diving, the rabbit struggled to escape. In both fear responses, heart rate was dramatically reduced, yet the responses appeared to be brought about differently. In passive fear, both portions of the autonomic nervous system were involved; with forced diving, the parasympathetic portion alone was responsible for the response. Maybe speed of onset was of less importance during forced diving than in the passive fear response? The picture appeared less focused than ever. More data were needed.

It appeared I had made a grave error in judgment. In my eagerness to investigate forced and natural diving I had overlooked the obvious. How could I observe natural dives with telemetered swamp rabbits? Forced diving required no imagination. I simply placed the rabbits inside a wire cage and submerged them. That was easy and the results publishable, but my major goal was to compare forced dives with natural, voluntary dives. My one observation of a single wild swamp rabbit diving did not an experiment make. Field studies would be extremely difficult. Buzzard Island would not hold a frightened swamp rabbit, for it would swim ashore and be gone forever. My chances of observing natural diving would require weeks or months of careful observation and night vision equipment to see in the dark. I had neither the time nor the money to invest. To publish the forced diving data alone would only add to the confusion already in the diving literature. I needed to come up with a method for studying voluntary diving under laboratory conditions. I needed to train wild swamp rabbits to dive under water in the laboratory. Training a dog was one thing, but training a wild animal appeared impossible. I consulted with an animal behaviorist.

I should mention here that I hate prejudice. The concept of judging someone by the language he speaks or by the color of his skin is simply wrong. It is also unscientific. I hate prejudice in others and am even more abhorrent when I see it in myself. The concept of judging a group of people from limited information gained by hearsay or by knowing one individual is equally loathsome. Yet zoologists are human and prejudice seems a part of humankind. Although I hate prejudice, especially in myself, there does remain one area I still harbor prejudice. It is not prejudice toward race or ethnicity, but toward people of one particular science discipline. I confess I hold prejudice against animal behaviorists in this country. Animal behaviorists in Europe I respect. They often study animal behavior outdoors under natural conditions. In marked contrast, American animal behaviorists often work with laboratory animals in unnatural Skinner boxes. I have several problems with this approach. First, laboratory animals are at best artificial. They have been selected for docility and seldom display behavior even remotely akin to that of their wild counterparts. But there is an even more important concern.

Behaviorists (or psychologists, as most prefer to be called) are far too quick to apply the results of how a few rats learn something in Skinner boxes to the way school children should be taught. Few seem to use rigorous statistical analysis in looking at their experimental results. Psychology teachers and students I knew added to my prejudice. Allow me to relate one amusing anecdote involving a psychology professor and psychology student.

While working with rattlesnakes at Southwestern Oklahoma State University, we had an old but large laboratory in which we housed about 100 live rattlesnakes. We also maintained colonies of laboratory rats and mice to provide food along with an ever-changing assortment of other wild animals as well. As one might expect, students and some faculty referred to our lab as "the zoo." It remained unclear whether they were referring to the animal residents or to those of us who worked there. A psychology professor, who shall for obvious reasons remain unnamed, visited our laboratory and asked Dr. Landreth if he would allow one of his students to use our facility. He had a student (let's call her "Mary") who wanted to study the effects of "positive energy" on a wild animal. Amused and curious, Dr. Landreth agreed. Mary wanted to work with a field-caught Florida pack rat. This is one of the largest and most aggressive rats in western Oklahoma. They were abundant. Mary's goal was to come in each day and give off "positive energy" by thinking nice thoughts and saying nice things to the rat. The final test would come at the end of the semester, when she would reach in and grab the rat to see if all the "positive energy" had tamed it.

The next week we caught a large, vicious Florida pack rat and introduced it to our largest rat cage. Mary faithfully came in every day, sat in front of the cage, and talked soothingly to the captive. The entire room was filled with kind thoughts. "Positive energy" thus flowed two hours a day for the entire semester and was directed toward the fearful pack rat. My respect for psychology as a science discipline reached new depths. With the end of the semester, the moment of truth came. Mary, her psychology Professor, Dr. Landreth and the rest of us even remotely related to the laboratory were present for the momentous event. Much to our surprise, Mary remained true to her proposal. No doubt her grade hung in the balance. Without fanfare she opened the cage, reached for the pack rat, and grabbed it. One does not need a science background to predict the results. The frightened rat turned and bit the hand that held it again and again. Mary flung the bewildered rat down, swore loudly and left. So much for positive energy and pack rats. I tell this story to help you understand how difficult it was for me to keep an open mind as I approached our resident animal behaviorist for help with my research.

Yet help he did. I outlined my dilemma of needing to train wild swamp rabbits to dive in the laboratory. In less than ten minutes he offered a plan and a student assistant. I accepted. The plan was simple. We would modify a large aquarium as a laboratory diving test chamber. Training would consist of placing a hungry swamp rabbit on a perch at one end of the nearly empty aquarium and food on a perch at the other end of the aquarium. A partial partition was placed in the center of the aquarium separating the two perches. In order to find food, the rabbit-in-training had to duck under the partition. With each training session, more water would be added until the rabbit would have to dive underwater to reach the food. What a simple plan, yet I honesty do not think I would have ever thought of it without the help of my colleague in psychology. This again showed the value of multidisciplinary consultation…even with psychologists. My prejudice was replaced by respect.

The diving chamber was quickly built and rabbit training commenced. The psychology student was competent and excited about applying her behavioral training to research, and she was getting a grade for her efforts. I was pleasantly surprised at how quickly the rabbits learned to traverse the tank for food. The only concern was the rabbits seemed apprehensive by the presence of the investigator or trainer. This was quickly remedied by closed circuit TV. The sound channel was used to record the telemetry signal. The next chapter in my research of passive fear was about to be written.

For this portion of the study I wanted to separate the fear component from diving. I had ample data from forced diving and needed to record voluntary dives for a food reward. Training required several weeks, adding water depth with each training session. Food was withheld for 24 hours before each training session, and the rabbits were trained twice a week, with two more inches of water added each time. After a few weeks, each hungry rabbit would immediately leap from the perch, dive under the partition and find the food. The entire event was video taped along with the telemetered heart rate. Soon I was awash with voluntary diving data. The results were exactly as I had predicted; yet still exciting. During voluntary diving, swamp rabbits did not slow their hearts. The rabbits initiated the dives voluntarily. Hunger had replaced fear as a driving force. This experiment provided more evidence that fear was involved during the forced dive. It also added credibility to the earlier alligator studies at Big Lake. Some of the puzzle pieces were fitting together nicely. A picture of passive fear was taking shape…a picture no other person on earth had seen. Research was again like climbing a mountain no one had climbed. I was getting yet another grand view of nature seen for the very first time by any human. It felt good.

# 7

# The hiding place

*Science is built up of facts, as a house is built of stones; but an accumulation of facts is no more a science than a heap of stones is a house.*

—**Henri Poincare**

Snow was falling gently as I looked out the dorm window at the courtyard of Balliol College. Bright yellow dandelions dotted a carpet of green, betraying the tardiness of the fleeting March snow. Balliol College, founded in 1263 AD is one of the two dozen colleges that make up Oxford University. Oxford University is old—really old—older than my home state of Oklahoma—older by centuries than the United States. Classes were being held at Oxford University before Columbus crossed the Atlantic. My mind has difficulty comprehending a man-made institution lasting eight hundred years, yet Oxford University had its beginnings before 1200 AD. Balliol College is over 700 years old. The dormitory my students and I were staying in had been in use over 400 years.

There were few indicators of the awesome age of the buildings. The old structures were well maintained and, unlike old buildings in the United States, showed no signs of graffiti, or cigarette burns, or broken glass, or loose banisters. Aside from the dated cornerstone, the most striking evidence of age was found in the marble stairs. Hundreds of years of student foot traffic had eroded half-inch deep shoe tracks in each marble step. The time and number of steps required for shoe leather to wear away marble boggles the mind.

I had been invited to Oxford University as keynote speaker for an International Radio Telemetry conference. I brought three research students from Northeastern Oklahoma State University: Lisa Causby, Janie Worth and Kevin Dawes. It was the first time any of us had visited England. We arrived in London late at night and the train ride to Oxford took an eternity. Once in our rooms, we collapsed for 14 hours. In spite of our excitement, jet lag and biological clocks took their toll.

The conference was well attended and inspiring. The research paper sessions were exhilarating, but my greatest thrill was meeting the movers and shakers of modern radio telemetry. I was meeting the people whose technical papers I had followed—people from all over the world. Conferees hailed from Africa, Russia, Australia, all over Europe, the United States and Canada. Most spoke fluent English; some were hardly intelligible; all were entertaining. I still had trouble believing I was actually at such a prestigious conference and that I had been invited to speak because of my alligator telemetry work. The only other keynote speaker was Dr. Tom Fryer from NASA. We represented two extremes of the telemetry world. Dr. Fryer had virtually unlimited telemetry budgets. Human life and world opinion of the United States depended on his measurements. He used the most advanced telemetry in the world. My telemetry, on the other hand, might disappear after the animal's release. My approach was to develop "disposable" multichannel telemetry transmitters costing less than twenty dollars to build. It was a humbling, yet exhilarating experience.

When the time finally came for me to give my keynote address, I was so frightened I could hardly breathe, swallow or speak. Fortunately the large podium kept my shaking knees hidden from view. It was a hundred times more terrifying than jumping on an 8-foot alligator. Once I got into the presentation I relaxed, but I doubt I will ever again be so horrified. Two of the three students that accompanied me presented technical papers at the meeting. One student, Janie Worth, was offered a post-doctoral position on the spot, but had to decline. Lisa Causby was offered a graduate position and also had to decline. Both were only college sophomores. I was pleased that they were so well prepared as to be confused with college graduates. They had worked long and hard for their moment in the sun.

Over one hundred scientists attended the conference. I met nearly all of them and have corresponded with two dozen. Three professional relationships developed into friendships that have survived the test of time. Charles Amlaner, then completing a post-doctoral position at Oxford, organized the conference. He and his wife were from the United States and they were especially nice to my students and me. Shortly after the conference he returned to the United States for a university position on the West Coast. He then served as the Chairman of the Department of Zoology at the University of Arkansas. We have remained in touch and he became the president of the International Radio Telemetry Society. He organized another international conference in 1989 in Arkansas, and I again participated.

A second long-term friendship fledged at the Oxford conference was with (the previously mentioned) Geir Gabrielsen. A Norwegian, he was completing his doctorate and investigating the heart rate response of Arctic animals to fear. He had come to many of the same conclusions as I and was using some of the telemetry devices I developed. We talked for hours and an immediate camaraderie was born. More about him and his work on the pages that follow.

The third lasting bond made at Oxford was with Craig Johnson. He was from Wales and was completing his undergraduate degree in London. Craig was looking for adventure. He singled out three conference presenters and wrote letters of inquiry to each regarding possible co-research. My response brought Craig to Oklahoma for a summer research project. He was our houseguest for the summer and fit into the family like my own kin.

Craig was an excellent observer, a hard worker, and an outstanding nature photographer. Following our summer together he and his wife traveled to Northern India and observed wild tigers for a year. Before departing India, he had an audience with the former Prime Minister of India, Indira Gandhi. They discussed wildlife and she was environmentally aware but limited in what she could accomplish with the available resources. Today Craig works for the World Wildlife Fund.

Together Craig and I decided to study the fear response of squirrels and chipmunk. These were common eastern Oklahoma animals and Buzzard Island would again provide excellent habitat. As a boy, I hunted squirrels and was familiar with their behavioral response to disturbance. If frightened on the ground, they would run to the nearest tree. Once in the safety of a tree they had several options. Sometimes they would leap from the branches of one tree to another and travel a considerable distance far above ground. This was often the case when they were some distance from their nest. Most of the time they would hide in the first tree they climbed by moving to the opposite side of the trunk or crouching low between two branches. I wanted to monitor their heart rate during the hiding episode.

There were no heart rate data available for any species of free-ranging squirrels. Robert Ruff, a graduate student at Utah State University, accidentally discovered a sudden slowing of the heart on a telemetered Uinta ground squirrel. Robert was trying to capture a telemetered ground squirrel, and its heart slowed while it was hiding in the undercarriage of a car. This observation, plus the behavioral response of squirrels creeping around the opposite side of a tree, made squirrels logical candidates for the next study.

Squirrels are plentiful in eastern Oklahoma, yet they proved difficult to capture. Their build is long; their movement agile. We tried to use the same type of box traps we used effectively to trap cottontail rabbits and woodchuck. The squirrels would enter the trap for the corn bait, but could leap to safety before the trap door closed. Special, longer traps were purchased, but still the squirrels prevailed. Finally, after talking to some old trappers, we learned a simple technique. We closed one end of the trap. When the treadle tripped and the box closed, the squirrel was facing the wrong way, and could not turn around and leap before the door closed. Success was sweet.

Several gray squirrels and fox squirrels were captured and heart rate transmitters attached. As usual, preliminary measurements in the laboratory were taken. This was necessary to assure the equipment was working before releasing the animal, and some useful information was sometimes collected. The preliminary data helped in the design of field experiments.

The laboratory data from squirrels was confusing. When the investigator or dog approached gray squirrels, they always retreated to the safety of a nest box. Heart rate slowed about 25 percent. Fox squirrels responded to the same stimulus by trying to escape. Their heart rates increased 160 percent. Without the previous differences in the heart rate response of caged and free ranging alligators, I would have been discouraged. Instead, I assumed the behavioral and physiological responses under more natural conditions in the field would tell a different story.

Undisturbed Gray Squirrel

The squirrels were released on Buzzard Island at Lake Tenkiller and left alone for a week. The transmitters were intermittently checked from the boat and were working exceedingly well. Since the squirrels spent much of their time high in large trees, the transmitter range was phenomenal. At times the signals could be heard more than five miles from the island. In addition to checking out the transmitters, Craig spent considerable time on the island. He constructed an observation blind on the ground and one high in a tree. In his leisure time he took a series of beautiful photographs of the habitat. A good nature photographer finds beauty everywhere. Several of his photographs have been published. I can think of no better way for him to remember my country.

A second research student, Kelly Martin, also assisted in the squirrel study. She and Craig shared interests in the natural beauty of the great outdoors. Craig captured beauty with a camera; Kelly with pen and ink sketches. Craig and Kelly were anxious to begin the study. Both spent several days reading about telemetry and squirrels and together helped design the study. The time to get actual field results was drawing near and we were excited.

The first experiment began with Craig perched high in his tree blind. Kelly remained hidden in some limestone outcroppings. I was operating the telemetry equipment, recording data and taking notes. A telemetered gray squirrel was in sight feeding on the ground. The telemetry signal was loud and clear. We were in communication via walkie-talkie. On Craig's command, Kelly started approaching the squirrel. It was unaware of our presence. Its heart rate was erratic, but correlated with behavior. When running on the ground its heart raced; when it stopped to investigate something its heart slowed. Kelly heard the squirrel long before seeing it. Although the cover was dense, it was in a small clearing. She stepped on a dry branch. It snapped, and the squirrel immediately stood on its rear feet and saw her. Without hesitation it bounded about ten yards to the nearest tree. Its heart rate doubled from about 200 beats per minute to over 400 beats per minute within a second. Once it reached the safety of the tree, it hid from her view on the far side and its heart immediately slowed. Craig and I witnessed the entire episode unseen by the squirrel. It remained hidden from Kelly with a very slow heart rate. Success! Once again a frightened wild animal slowed its heart while hiding. Similar results were obtained with other squirrels during the next few weeks.

The most significant finding from the squirrel studies was not that their hearts slowed while hiding. This had become what I expected. Yes, it was nice having two more species showing the same response to threat, but there was more. The differences between laboratory and field results were striking for the fox squirrel. In the laboratory, threat resulted in attempts to escape and increased heart rate. The exact opposite cardiovascular response was observed under natural conditions. In the field the frightened squirrel displayed the passive fear response by hiding and slowing its heart rate.

This had profound implications for medical research. Previously, nearly all medical research was conducted using traditional laboratory animals. That is changing. A growing number of wild animals are now used as models for a variety of diseases and to model various physiological functions. Few medical researchers have a zoological background. Fewer still have field experience. My point is obvious. Wild animals maintained in laboratories, no matter how healthy, often do not have the same behavioral and physiological responses as they do under natural conditions. The fox squirrel's heart increased when approached by the investigator in the laboratory. Under natural conditions the same stimulus resulted in passive fear, hiding and reduced heart rate. This is an important point as more and more wild animals are being used as models for disease and physiology.

Eastern chipmunk

The next study the three of us embarked on involved eastern chipmunks. A chipmunk was trapped and the telemetry device attached. Preliminary laboratory results were intriguing. The chipmunk was housed in a large aquarium. Soil in the bottom allowed the animal to build shallow burrows. A small cardboard box provided a safe haven. The animal was tested in the laboratory by the approach of the investigator. It responded behaviorally by retreating to the safety of the cardboard house. Heart rate dropped from 259 to 184 beats per minute when it retreated to the safe hiding place.

The previous work with swamp rabbits revealed the importance of both portions of the autonomic nervous system for the fear response. We wanted to determine the importance of the parasympathetic nervous system by again treating the animal with atropine. To rule out the effect of handling and the injection, we decided to do a control experiment first and inject saline as a placebo.

The only way we could catch the chipmunk was to trap it inside the cardboard box. But once we did this, the chipmunk avoided the cardboard house. With this change in behavior came a change in the physiological response. The chipmunk no longer had a safe hiding place and tried to flee with each approach of the investigator. Heart rate increased with attempts to flee. This seemingly trivial experiment provided another piece to the puzzle. It was not enough to have a physical place to hide. The frightened animal must consider it a safe hiding place. Similar experiments were repeated with additional animals with identical results.

It is unlikely that squirrels, chipmunks and rabbits were trying to conserve oxygen, yet they showed the same response as alligators and woodchucks. The response occurred while hiding from danger. Could the passive fear response somehow enhance hiding? Could a reduction of heart rate and breathing help the animal remain invisible to the predator? It was just a thought, but a thought I could not dismiss. I still had an incomplete puzzle, but some parts of the picture were coming into better focus. The passive defense response to fear was more widespread and more complex than I had imagined. From my studies and the work of other investigators scattered throughout 50 years of scientific literature, the response was seen in fish, amphibians, reptiles, birds, and a variety of mammals. Animals living in water, on the ground, underground and in trees showed the response. It was not enough to be frightened; a safe hiding place was also necessary. The degree of passive fear, at least as measured by heart rate, corresponded to the intensity of the stimulus. Passive fear reduced oxygen consumption, even for animals that appeared to not need it. Parts of the puzzle were beginning to make sense. Data from other animal species occupying different ecological niches were needed.

# 8

# Nature's mobile home

*That is the essence of science; ask an impertinent question and you are on the way to a pertinent answer.*

—Jacob Bronowski

Lazy days of summer

Throughout the history of modern science, experimental design has been the topic of endless discussion, thousands of scholarly papers and myriads of books. It seems to change with each new generation of scientists. If ten scientists, competent in their respective disciplines, were assigned the task of defining good experimental design, there would be at least ten diverse definitions (and possibly more).

I have heard university professors say with furor, "An experiment without statistical analysis is not science." Nonsense. Watson and Crick were awarded the Nobel Prize for their work unlocking the DNA code, yet used no statistics. Other scholars adamantly proclaim that an experiment without a control is useless. Again major scientific breakthroughs have been made without carefully constructed control experiments. Science is immensely broad, and different scientists working in the various disciplines use dissimilar methods. To me, experimental design is simply asking nature a question; and, as in language, there are many correct ways to ask a question.

Even within zoology, few scientists agree on the best method of experimental design or the all-important task of how to select the proper animal model for study. Purists wax long and eloquent about using the most appropriate animal regardless of where it is found. The one that best demonstrates the physiological feature to be studied is touted as the best choice. For example, if one wanted to study the physiology of deep diving, then the largest whale should be used, for it is they that dive deepest and longest. Here realism rears its ugly head. Even purists have budgetary constraints and can not always study the animal of choice. Furthermore, many species are classified as endangered and are thus beyond reach for experimentation.

My approach to research has always been pragmatic; I work with whatever is available. One of the members of my doctoral committee remarked that if I were serving a life sentence in prison with one foot nailed to the floor, I would publish research results regarding the lunar effects on the number of steps required to walk around the circle. I have never lacked ideas for research—only time and, occasionally, money. Over the years I have received phone calls and letters from investigators at major universities and medical schools who, having recently acquired expensive radio telemetry equipment, wanted me to give them research ideas for using the equipment. Apparently, unlimited research budgets, fancy degrees and prestigious research positions do not spawn creativity. I am reminded of the words of Albert Einstein, "Imagination is more important than knowledge." I agree. Perhaps it is imagination coupled with curiosity that makes a successful scientist…and let's not forget the importance of luck.

The box turtle, nature's mobile home

Selecting the next animal model to continue the passive fear investigation was easy. I had studied alligators that hide by diving underwater and woodchucks that scramble underground when frightened. Rabbits and squirrels hide by using whatever cover is available. Next I wanted to study an animal that carries a hiding place with it. I chose the path of least resistance and studied an old friend. Box turtles would be next.

It has been said that turtles, like humans, can accomplish nothing without first sticking their necks out. They, like rabbits and woodchucks, have influenced American folklore. An entire generation of kids went crazy over Teenage Mutant Ninja Turtles. My infatuation with box turtles began long before turtles were popular crime fighters. In order to protect her prize-winning tomatoes from being eaten, my grandmother would bring turtles to me from the garden. They were better than any toy, as they needed no winding or batteries. Occasionally, my grandfather would even bring turtle eggs he found while plowing the fields. We kept the eggs warm and moist, and released the baby turtles upon hatching.

I played with box turtles for hours, and even as a child was fascinated with their response to fear. When a dog (or mischievous boy) pestered them, the box turtles retreated inside their shells and closed the hinged door. Dogs could find nothing to bite and would give up in frustration after barking awhile. The box turtles were safe even from the most vicious dog (or small boy). By accident, I discovered they would remain stationary for hours if placed in the moving shade of a tree blowing in the Oklahoma wind. They would slowly extend their necks and look first to one side then the other. If the shade moved, they would withdraw to the safety of their mobile home. Even as child-zoologist, accidental discoveries directed my path. All I needed was a highly motivated student interested in studying box turtles.

Martinho de Carvalho grew up in central Brazil. His family had no money, yet with the persistence typical of bright students in developing countries, he finished high school and secured a college scholarship from the I.N.P.A. Ecology Laboratory in Manaus, Brazil.

I had met Martinho during a brief visit to Manaus, Brazil, where he had served as translator, tour guide, and friend. I visited Manaus in the dry season, yet the Amazon River was two miles wide and roaring. I found it difficult to believe that the river must travel over a thousand miles to reach the Atlantic Ocean. There are twelve tributary rivers that empty into the Amazon River, each larger than our Mississippi River. More species of fish inhabit the Amazon basin than are found in all other fresh water in the world.

Martinho was involved in a Manatee conservation program, and even let me bottle-feed a young calf whose mother had been killed by poachers. The group had rescued dozens of young Manatees from slaughter. The adults are still hunted for blubber on the Amazon River. It is one thing for wildlife conservationists around the world to pass laws protecting endangered species, but quite another to convince a hungry man he should not kill animals he has always hunted to support his family. Complex problems do not have simple solutions.

By our standards, funding for this crucial work was non-existent, yet many eminently qualified scientists were devoting their careers to the important work of manatee conservation. Martinho wanted to learn more about electronics and telemetry. Soon after my return home he secured travel funds to visit my laboratory and spent four weeks as our houseguest. He immediately took to reptiles like a veteran herpetologist, and since I had wanted to investigate the heart rate response of frightened box turtles for years, they were a logical choice for study during his short research visit.

Following a week of library work and field observation, we collected seven ornate box turtles and attached telemetry devices. A variety of stimuli were used, including approach and contact by investigator and dog, moving shadows, forced submergence and voluntary dives. Box turtles, like every other species tested, slowed their hearts when approached by the investigator. Behaviorally they pulled their legs, tails and heads into their shells and closed the hinged door. There was no need to look for cover or retreat to a burrow. They carried their own hiding place with them.

Box turtle hiding in safety

Like the response of woodchuck, the magnitude of the response was related to the intensity of the stimulus. A passing shadow produced a small, but significant reduction of heart rate. Mean heart rate reduction from the visual approach of the investigator or dog was 37%. Touch by the investigator resulted in a marked 68% reduction in heart rate. In each case fear bradycardia was immediate and required less than two seconds to fully manifest itself.

Forced diving resulted in a 44% reduction of heart rate but the onset was slower as I had seen in swamp rabbits. Sometimes over a minute and a half were required for the diving bradycardia to develop. Such slow onset bradycardia suggested chemical sensors within the animal were responding to increased carbon dioxide levels as the dive persisted. That was to be expected from the diving literature. As was true with alligators and swamp rabbits, voluntary dives occurred without any significant change in heart rate. It was apparent that diving animals made two kinds of dives. Voluntary dives, which could be terminated at any time, involved little or no change in heart rate. Forced dives, on the other hand, represented a real threat since the animal had no control over the onset or termination of the dive.

The box turtle was an animal far removed taxonomically and ecologically from any species previously tested, yet it showed an identical response. The widespread nature of the response was becoming clear. It was intriguing that such a response had gone unnoticed by other investigators. But it seems we often see only what we expect to see. How many other responses had been missed simply because scientists were not looking for them?

An intriguing aspect of scientific research is that surprises can occur from the most unexpected places. Consider the lowly turtle, an animal that is literally underfoot. It is easy to capture and handle; yet no one had recorded its cardiovascular response to fear. Knowledge of one animal increases respect for all living things.

Nearly ten years had passed since that accidental discovery at Welder, yet I was no closer to knowing the real value of the response to the animal. Since so many animals displayed the response, it must be important. What is its adaptive value? How does it help an animal survive? Does it occur in people? These and a host of related questions haunted me day and night. How many more pieces to the puzzle must be collected before a picture emerges? My worst fear was that I already had enough pieces of the puzzle, but was failing to recognize what the data were trying to tell me. What was I overlooking?

# 9

# Far, far away on a tropical isle

*Science is dirty work, but someone has to do it.*

—Anonymous

Zoologists enjoy vacations, just like normal people. To the trained observer however, there are subtle differences. Consider, for example, the way vacation destinations are selected. Most people decide what activity they want to do on vacation and then choose the best place to do it. The strategy is the same whether the desired activity involves camping beneath a pristine forest canopy, collecting seashells on an endless beach or taking in the sights of a bustling city.

The zoologist's approach is somewhat different. To a zoologist, a vacation is a socially acceptable excuse to observe some new exotic animal in its natural habitat. Hours of reflection are given to the selection of the object of study. Once the perfect animal is chosen, a quick search of the scientific literature yields the animal's preferred habitat and what studies have already been published. This directs the vacationing zoologist to any obvious research remaining to be done. Research plans then commence. Maps are procured and the vacation spot selected. Next, as with any family vacation, funding must be sought. For the vacationing zoologist, research funding helps pay for the necessary equipment, publication of the results and more. Funding also adds credibility to the "working vacation" argument used to appease suspicious friends and coworkers. Sometimes research assistants are required for the timely completion of the study. Other zoologists often make the necessary sacrifice to help a fellow zoologist in need and to study a new animal in an exotic place.

In addition to my study of passive fear, I continued to investigate reptilian thermoregulation. One summer at Welder, a fellow graduate student happened to catch a near record-size spiny soft-shelled turtle on a fishing line. He knew of my interest in reptile thermoregulation and offered it to me. At the time, I was doing some simple heating and cooling experiments with other reptiles and decided to take advantage of the catch. Unlike most other turtles, soft-shelled turtles have a slippery leather-like skin. They are predominantly aquatic animals and excellent swimmers. They are coin-shaped and thin with a relatively high surface-to-volume ratio, so I did not expect them to be particularly good thermoregulators. Poor thermoregulators heat and cool at roughly the same rate, which differs only slightly from the response of dead animals, which must heat and cool at exactly the same rate. Good thermoregulators heat faster than they cool, due largely to increased blood flow during warming. Such simple experiments had been preformed on a wide variety of reptiles for over two decades.

Soft-shelled turtles had not been tested, nor was there any indication in the literature that they spent much time basking. Based on the literature and their shape, I was not expecting the experimental results to be very exciting, but publishable nevertheless. Much to my surprise, the large turtle heated more than twice as fast as it cooled. It was apparently able to heat so rapidly by increasing blood flow to its leathery shell. Small blood vessels became visible during warming and the heart beat 5.5 times faster while warming than it did at the same temperature during cooling.

Upon returning to Oklahoma, I collected several more spiny soft-shelled turtles of different sizes, tested them and published the results. At the time the results were published, soft-shelled turtles were the best reptilian thermoregulators studied. They also showed the largest difference between warming and cooling heart rate. To me this indicated that heart rate was controlled more by thermoregulatory requirements than metabolic requirements. This was important because physiologists had been taught for decades that the primary driving force for the cardiovascular system was the maintenance of blood pressure to meet tissue metabolic requirements. All other aspects of circulation were thought to be secondary. The heart rates of animals I had studied exhibiting passive fear slowed quickly and dramatically. It is difficult to imagine their overall metabolism could slow so quickly. The soft-shelled turtle's heart seemed more important as a heat pump than as a provider of oxygen and remover of carbon dioxide. More data were needed from a larger turtle.

My first choice was the giant leatherback sea turtle. Unfortunately, they were difficult to keep in captivity, were hard to capture, and, because they were an endangered species, it was very difficult to obtain permits for research. Another large sea turtle, the green sea turtle, was more accessible (although it, too, was endangered). Captive turtles were available at the Cayman Island Turtle Farm. I contacted the director, Dr. Jim Wood, and he was delighted by my interest and agreed to provide the facilities and turtles. Only two more things were required. My vacation plans were nearly complete. I needed only to secure funding and find a willing assistant.

Funding required over a year to secure. After several unsuccessful attempts, I approached the British government for support. Since the British government heavily subsidized the turtle farm, they were willing to assist with some of the research expenses and to provide housing for me and one assistant. All that remained was finding that assistant.

Several colleagues were considered, but most had teaching commitments and could not get away. I was at that time writing a research paper with a college student, Nancy Long, from Oberlin College in Ohio. We had met several years earlier when I gave a seminar at the Oklahoma University field station at Lake Texoma. I had lectured about alligator thermoregulation and the passive fear response. At the time, she was a visiting high school student taking college courses during the summer session. Two years later, she wrote to me about doing some lizard research over Christmas break. She came as a houseguest and worked with Dr. Stanley Robertson and me. Most of her time was spent at the Oklahoma University library, and we began to work on a review paper about lizard thermoregulation. She was a hard worker, loved reptiles, and enjoyed travel. The following year I mentioned the possibility of studying sea turtles at the Cayman Island turtle farm. Not surprisingly, she was willing to make the sacrifice. Since I had heard that scuba diving off the island was considered the best in the world, I decided to become a certified scuba diver before the trip. Nancy agreed to do the same. Vacations require detailed planning and careful preparation.

Our preparation for the trip included an extensive literature search to find everything written about sea turtles. For over a year I studied sea turtles. I knew about their reproduction, diet, growth rates and basking behavior. I knew of their ocean migrations to the beach of their birth. And I understood the perils faced by hatchling turtles during their race to the sea and beyond. I was anxious to work with sea turtles first hand.

Attempting a publishable scientific study in two weeks under unknown conditions is not a trivial logistic exercise. I contacted my travel agent to determine the largest size of luggage I could take without paying a penalty. Then I purchased a trunk of those dimensions and designed a portable wind tunnel. When disassembled, the entire wind tunnel and all the necessary electronic and biological equipment for the study could be stowed inside the trunk. I practiced assembling and re-packing the equipment to make certain I had all the necessary tools. Since I wanted to record heart rate and knew I would not have access to ECG recorders, I designed a small portable device that would convert turtle heart rate to an audible tone that could be recorded on a small cassette recorder.

After months of library work and careful planning, Nancy and I were finally ready to meet our first sea turtle. We arranged our flights so we could meet in Miami and visit some of my colleagues at the Everglades National Park. After touring the park and swapping a few alligator tales, it was on to a tropical paradise for a zoologist's working vacation.

The Cayman Islands are located about 150 miles south of Cuba and provide a playground for the rich. Tourists outnumber the native population during the busy winter months. Palm trees line miles of sandy beaches. Scores of diving boats carry sightseers over the clear azure water. Underwater sights include reefs, sunken ships and more. It is a diver's paradise, with visibility often exceeding 200 feet. Coral, giant basket sponges and colorful fish abound. In short, it was a tropical paradise. But I was ready to meet the green sea turtle.

Occasionally, at a party or other social event, you meet a special person and everything just clicks. You are both interested in the same hobbies and music, you know some of the same people and have read the same books. You understand each other in an extraordinary way and anticipate what the other is going to say. Although you have just met, you feel you have always known each other. In California parlance you have just met your soul mate. This happens to most of us only once or twice in a lifetime. It has happened to me only once, but I will forever savor the moment. The place was Grand Cayman Island and my soul mate was a green sea turtle.

Bonding with a friend on Grand Cayman Island

The Cayman Island turtle farm had thousands of turtles. After a brief tour of the facility, Nancy and I were introduced to our first sea turtle. The turtle was beautiful; the experience unforgettable. I held it and turned it over. I studied its face and shell and feet. I savored the feel and smell of its skin. Its movements were graceful; its beauty breathtaking. I was overwhelmed with emotion. It was an exhilarating experience. In my hands was the creature I had been thinking about for so long. It was like seeing a long lost friend.

The study was successful and we discovered green sea turtles are capable of heating in a smaller fraction of their cooling time than any reptile ever studied. We completed two separate turtle studies and still found time to explore the beaches and to dive the reefs. The water was clear and the diving fantastic, but it could not compare to the emotion of meeting my first sea turtle.

Although the sea turtles did not exhibit the passive fear response under our highly artificial experimental conditions, there were important discoveries. One of the exciting things we discovered was the complete uncoupling between body temperature and heart rate of cooling the sea turtles. They warmed in 41 percent of the time required to cool, making them the best thermoregulator ever described. More importantly, they exhibited the largest difference in warming and cooling heart rate. Warming heart rate exceeded cooling heart rate by a factor of 7.2. The most unusual discovery was that heart rate immediately dropped to about 10 beats per minute when the animals were placed in cool water. Heart rate remained slow and constant throughout the cooling episode. It did not show temperature dependency. This was important because tissue metabolism is always related to temperature. At higher temperatures all tissue requires more oxygen. Here again was strong evidence that the heart has other even more important functions. For the green sea turtle, reducing blood flow and thereby slowing the rate of cooling appeared to be more important than providing oxygen to its tissue. Although not directly related to passive fear, the Cayman Island turtle study indicated oxygen requirements were sometimes overridden by other factors. And it was a delightful way to spend two weeks vacationing in January.

# 10

# Mountain boomers in hiding

*Four things are small on the earth, but are exceedingly wise: the ants are not a strong folk, but they prepare their food in the summer; the badgers are not mighty folk, yet they make their houses in the rocks; the locusts have no king, yet all of them go out in ranks; the lizard you may grasp with the hands, yet it is in kings' palaces.*

—King Solomon

Mountain boomer basking in the sun

From my earliest recollections, reptiles held a special fascination for me. As a child, I thought their movements more graceful than mammals', their colors brighter than birds'…and I could catch them. But liking reptiles has not been popular since the Garden of Eden. I was ridiculed as a child for liking reptiles better than mammals, and at times was even snubbed as a graduate student.

Within academia, herpetologists are often shunned and belittled as "non-scientific snake grabbers" or worse. Even today, my best friend does not comprehend why an otherwise normal middle-aged college professor would suddenly stop the car and excitedly jump out to grab a snake crossing the road.

In contrast, bird watching is a socially acceptable form of recreational zoology. Ornithologists have a level of community and academic respect unattainable by herpetologists. I like birds and on several occasions even attempted group bird watching. I really tried. It just didn't work for me. Snakes and lizards are best caught in the late afternoon or early evening. Bird watchers arise well before the sun and tramp through dew-covered woods and fields, while any self-respecting reptile is still curled up in a hole. My fellow bird watchers were perpetually disenchanted with me. With eyes trained skyward, they couldn't comprehend why I was always the last person to see yet another red-breasted nuthatch. I never had the courage to explain that my eyes too were looking for movement, but on the ground. It has always been impossible for me to pass a well-rotted log without rolling it over to see what slithers out. Bird watchers would just shake their heads and leave me behind. Bird watchers never return to the car with their quarry in hand. But my typical bird watching expedition ended with my photographing and releasing a half dozen lizards, snakes, salamanders, and other assorted critters.

Western Oklahoma provides an excellent habitat for a variety of interesting lizards. Horned lizards abound in pastures and along seldom-used country roads. Children of all ages and out-of-state visitors find them entertaining. Box turtles and horned lizards are among the most popular show-and-tell topics at school as the late spring temperature climbs toward triple digits. Horned lizards are active throughout the hottest days and have an unusual thermoregulatory behavior. At the end of the day, as the sun sets and the temperature drops, horned lizards bury themselves a few inches below the surface of the sandy soil. As the morning sun begins to heat the sand, they shuffle up to just below the warming surface sand. When it gets warmer they push their heads above the sand. Warming rays of the summer sun heat their heads, and inside their heads are vast blood-filled sinuses. The blood is warmed by the sun and soon warms the entire body. Once they are warm and coordinated enough to flee if necessary, they emerge and look for food. If horned lizards become cool during the day, they orient their thin flat bodies perpendicular to the sun to take advantage of radiant heating. In the hot afternoons, they face directly toward the sun (with eyes closed), thus exposing the smallest possible surface area to the sun.

Ants are their favorite delicacy. Dens of large red ants punctuate the grasslands of western Oklahoma like periods on a printed page. (These large red ants are not to be confused with the more recently arrived fire ants.) As a boy, I would often catch horned lizards, only to release them at the nearest ant den to watch them eat.

But the largest and most colorful lizard in western Oklahoma is the "mountain boomer" or collared lizard. Occasionally reaching a length of 14 inches, it is the official state lizard. It is uncertain how they acquired the name "mountain boomer." Perhaps early settlers attributed the loud booming sounds of courting prairie chickens to these large, active lizards. They are aggressive and run fast. While running, they rise up and run on their large hind legs, with the long tail keeping balance. They eat a variety of smaller animals including grasshoppers, young birds, other lizards and small snakes. Early in the summer, males are bright green and yellow with a distinct black collar around their necks. Females are slightly smaller and show bright red spots in addition to green and yellow. Females lack the distinctive collars. Both sexes are difficult to catch because they quickly retreat to the protection of their burrow under a rock outcropping.

Like most other lizards, collared lizards mark their territory by doing a series of specie-specific bobs or push-ups. The effect of the display is to attract a mate, repel other males and to mark and defend their territory. Lizard push-ups provide much the same function as bird songs, except they are more exciting to behold. The characteristic collared lizard bob consists of two shallow bobs, pause, two shallow bobs, a short pause, followed by three deep bobs with the bright yellow dewlap expanded. It is elicited by the approach of another lizard, even a lizard of a different specie. As summer wears on, the bright colors of both sexes fade and the bobbing displays become less frequent. My passive fear research continued with an old friend, and a new one.

Many Brazilian college students dream of visiting the United States. They are very proud of their own country and do not want to move. They just want to visit. There is curiosity about life in the United States, and it is a status symbol to tell friends they have visited our country. While visiting Brazil to study passive fear in sloths and Timbu, I had the opportunity to lecture at several engineering colleges and medical schools. At the end of each lecture, college students always encircled me. A few asked questions regarding my research, but most wanted to exchange addresses and try out their English. They wanted a personal invitation to visit my home. One of those students was a young lady named Megumi.

Megumi was a Japanese Brazilian. There is a rich and varied Japanese culture in Brazil. There are Japanese speaking communities with Japanese public schools, Japanese art and Japanese restaurants. Cultural exchange is common between the two countries and Japan helps educate young Japanese Brazilian students in eastern culture.

Megumi liked reptiles, but could only visit for two weeks. Before she came, we decided she would study the heart rate response of mountain boomers to fear. I sent her several reprints to read and captured several large healthy specimens before her arrival. After introducing her to captive mountain boomers, we drove the roads and walked the canyons to acquaint her with the lizard in its natural habitat. She also needed to adapt to the 20–30 degree warmer climate. One of the most surprising aspects of my visiting the Amazon River was the discovery that temperatures are warmer here in Oklahoma. Shade temperatures in the rain forest seldom reach 85 degrees. Summer shade temperatures in Oklahoma often exceed 100 degrees.

Her second and third days were spent observing a population of collared lizards on a nearby hillside. Her observation skills were keen and her journal notes meticulous. She fell in love with mountain boomers, and observed 23 of the beautiful lizards for two days under fully natural conditions. I knew the neighbors who owned the hill. Although long-time friends of mine, they never understood why I would ask a foreign visitor to stay on their hill in the hot sun, watching lizards. They felt sorry for Megumi and brought her food and iced tea several times during both days.

With some book knowledge and two days of field observations, we were ready to begin the study. Several lizards were outfitted with heart rate transmitters and released in a large outdoor enclosure. Rather than construct a nearby blind, Megumi chose to remain hidden in a Chinese elm tree near the pen. Two days were spent collecting heart rate data and behavioral observations. In addition to heart rate data, she collected soil and air temperatures, wind velocity and relative humidity at 15-minute intervals. As expected, heart rate data was correlated with activity. Heart rate increased while the lizards were running or digging and was lowest during the cool early morning hours.

On the third day several disturbance experiments were conducted. Megumi was the observer (and data recorder) and I provided the stimulus. Fear induced running and running increased heart rate. Retreat to their burrows always resulted in a marked, sudden slowing of the heart. When I attempted to dig out the hiding lizard, its heart would slow even more. On one occasion, after digging out a lizard, it became aggressive and attempted to bite me. Its heart rate increased during this episode. In future experiments that particular lizard would not retreat to its burrow and did not slow its heart rate. Like the chipmunk, it no longer had a safe hiding place.

It was reassuring that each of these heart rate responses had been previously discovered with different species. Mountain boomer behavior in response to threat was most like that of the woodchuck, as was its heart rate response. Specifically both species showed increased heart rate as they retreated from the stimulus and ran toward their burrows. Both species showed a sudden marked slowing of their hearts upon reaching the safety of their holes. And both species slowed their heart rate even more in response to digging near the burrow. Finally, when capture of the hiding mountain boomer and chipmunk abolished the behavioral response of retreating to the burrow, the physiological response was altered as well. Certain pieces of the puzzle were fitting together nicely. A picture was emerging.

Several mountain boomers were released in a canyon on my farm, and data under natural conditions were collected. Again a variety of stimuli were used, including the approach of a dog, or me on foot, or riding the noisy family three-wheeler (an ATV). Each approach elicited the same response. The lizard would first rise up and look toward the stimulus. Next, it would run to within a few inches of its burrow and watch the approaching threat. Finally, the near approach of the stimulus would result in a rapid retreat inside the burrow. The first run-and-stop resulted in increased heart rate. The final retreat inside the burrow always resulted in a marked slowing of the heart. Digging or other disturbance near the burrow would achieve an additional lowering of the heart rate, just as I had discovered years before when my dog attempted to dig up a woodchuck safe in its borrow.

Perhaps the most rewarding aspect of the mountain boomer study was that there were no surprises. Part of this was due to my lifelong knowledge of the lizard. We had grown up together and shared many delightful hours. (No doubt more delightful for the child-zoologist than for the lizard). I had chased the great-grandparents of those same lizards I studied into some of the same holes as a boy. I had poked at them with a stick, wondering what they were thinking and doing out of my reach. I even remember on one occasion actually saying, "Mr. Mountain Boomer, why are you hiding? I won't hurt you. Please come out and play." Forty years later, with the help of a Brazilian zoologist and some sophisticated telemetry equipment, I continued that dialogue.

# 11

# Cat and mouse game

*But the beating grew louder, louder! I thought the heart must burst.
And now a new anxiety seized me—the sound would be heard by a
neighbor!*

—Edgar Allen Poe

With the notable exception of alligators, all the animals I had investigated were prey species. Even alligators slowed their hearts when they were hiding from a predator (man). Successful concealment from a predator could mean the difference between life and death for a prey species. But concealment of a predator from its prey is also important, and could mean the difference between eating or starving. I was becoming certain that the passive fear response augmented concealment.

The most convincing evidence was obtained by accident by an investigator in California. He was examining previously recorded heart rates of several fish and observed that at the same instant each fish stopped its heart. One fish stopped its heart for over 5 minutes. Gradually the hearts of each fish re-started and soon normal heart rates returned. What had caused the fish to stop their hearts at the same time? Weeks later, while examining the data, he realized a small earthquake had startled the fish and caused their hearts to stop. This was an odd twist to the passive fear response.

It had been known for years that certain predatory fish have a lateral line detection system so sensitive to electric fields that they can detect the ECG of another fish's heart. The beating heart could literally lead a predator to its prey. Stopping the heart is the only way the prey can avoid detection and death. This surprising discovery also proved there is more involved in the control of heart rate than the mundane metabolic needs of tissue for oxygen and carbon dioxide removal. It was becoming obvious how reduction in heart rate might at times be crucial to survival. What about other species living in different environments?

Although not predators in the usual sense of the word, ticks are known to follow the trail of exhaled carbon dioxide to a host. If a host reduced its metabolism and thus carbon dioxide production, the risk of being found would be reduced. Perhaps the passive defense response was an attempt to reduce all possible location clues of the hiding animal. Perhaps predators had hearing so sensitive they could hear the pounding heart of a nearby hiding animal, as fish could locate another fish by the electrocardiogram of its beating heart. Reduction of heart rate might help the prey avoid detection and thus enhance its chances of survival.

What about a stalking predator? What physiological adjustments occurred when a predator was stalking its prey? The predator had the same need to avoid detection.

A variety of predators were considered. I had a standing invitation with former Welder research student, Roy McBride to telemeter mountain lions in Great Bend National Park near El Paso, Texas. But the logistics just couldn't be worked out, so I had to choose another predator. I had a similar invitation to investigate cheetahs in South Africa. But my college refused to allow me to take the needed research time away from my teaching. Perhaps the solution to the problem lay at my feet, purring.

Domestic cats are excellent hunters. Every cat owner is familiar with the combination of patience, stealth, cunning, and explosive speed that leads to a successful catch. Cats of many species are capable of overpowering prey several times their size. Their agility, exquisite eyesight and hearing, retractable claws, strength, and sheer hunting skill have captivated the imagination of man throughout recorded history.

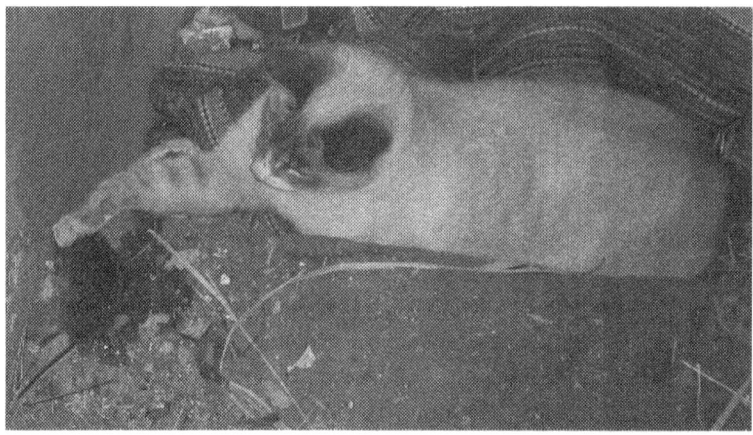

Cat and prey

Some of my fondest childhood memories involve dogs and cats. As a boy, my best friend was my dog, Tippy. He was part fox terrier and part unknown, but was always my faithful companion. We had two kinds of cats on the farm: barn cats and house cats. Both kinds earned their keep by protecting stored grain from rats and mice. My own children grew up with cats, dogs and alligators. One cat they were particularly fond of was Rocky. Rocky was a large, 14-pound black and white domestic short-haired cat. Although neutered, he was an excellent hunter and often brought rats and mice home.

The children did not enthusiastically receive the prospect of Rocky being used for research. They knew the risk of surgery was minimal, but were convinced shaving his tummy for surgery would humiliate him beyond forgiveness. What if he ran away in protest? Their fears subsided when I developed a strap-on harness for Rocky. Button ECG electrodes were held against his chest with an elastic harness so the humility of shaving was avoided. The small heart rate transmitter was sewed into the harness at the back. It took two weeks for him to become accustomed to the harness, and then it was time for Rocky to make his contribution to science.

The experimental procedure was conducted in the most natural of environments—our garage. We lived on a wooded hillside and often left the garage door open. An endless variety of native creatures were often found inside. It was Rocky's favorite hunting ground. The experimental prey was a laboratory rat. Since a frightened old rat could become a formidable foe, I decided to use young rats. I did not want the children's beloved pet to receive a painful bite.

The transmitter was switched on and pre-hunting or control data collected. The heart rate was slow and regular. Rocky felt honored to have me share his hunt. A young rat was released in view by pulling a string tied to the cage door. At once Rocky heard the release and saw the rat. The heart rate fell precipitously as the hunter dropped to the familiar stalking stance with belly on the floor and the tip of the tail swishing. The heart rate slowed further as he crept to within pouncing distance of the rat. The final attack resulted in an expected increase in heart rate.

I was at once awestruck. This was another piece of the puzzle I desperately needed. I had direct evidence of the adaptive value of the passive fear response. The response of this unlikely predator added credibility to the theory that the passive fear response made the hiding animal less obvious. Before this simple experiment was performed, all I had was speculation. Afterwards, I had hard evidence. The oxygen saved by the passive fear response was an added bonus for diving or burrowing animals. The cessation of movement, reduction of breathing and heart rate along with changes in blood flow appeared to reduce the chances of a prey species being discovered.

Predators also showed the response while stalking prey. Successful predators must avoid discovery by their prey. Anything that reduces the chance of detection helps the predator secure a meal. I was elated. It had taken me over ten years to discover the value of the response to the animals demonstrating the response. All the pieces suddenly fit together. That moment made the hundreds of experiments and years of effort worthwhile. Scientific discovery is its own reward for those who persevere.

I had answers for two of the three questions I set out to discover. The passive fear response was more widespread than anyone imagined. It appeared to be as widespread as the active defense response (sometimes called "fight or flight" response) and it involved many physiological parameters. Every species I had tested showed the response under the right conditions. That is, I could evoke passive fear if the animal had a safe hiding place or was a predator stalking prey. The adaptive significance suddenly appeared obvious. The response helped a hiding (or stalking) animal to remain undetected. The answer to the third question, relating passive fear to humans, came from a most unlikely candidate.

# 12

# Opossums and crib death

*Upon seeing the resurrection of Jesus Christ, the Roman guards shook with fear and became like dead men.*

—Saint Matthew

Progress in science is often best made by studying extreme cases. Investigating electrical current flow at a temperature of near absolute zero has enlarged our understanding of electrical current flow in general. Study of the effects of weightlessness in space has increased our knowledge of the physiological effects of gravity on earth. Animals that define the physical limits of survival increase our understanding of function under less severe conditions. Studying camels deepened our understanding of how the body copes with fever. These giant desert dwellers cannot afford the luxury of sweating or panting, for too much precious water would be lost. Instead, they store heat during the day and release it during cool nights. Each day they experience body temperature extremes that would cause other mammals to suffer heat stress or shiver. Sharks do quite well with levels of urea that would kill most animals. A better understanding of how sharks accomplish this might lead to less frequent dialysis treatment for kidney patients. The arctic ice fish winters in an unstable supercooled state and has taught us how other certain other animals adapt to extreme cold. To better understand the passive defense response to fear, I decided to study the animal with the most exaggerated passive fear response.

The passive behavioral response to fear that outranks all others is death feigning. The reaction to threat of our own native marsupial is legendary. "Playing 'possum" has become part of our language and culture. Death feigning is another one of those widespread, but seldom studied, behavioral responses. Many insects drop to the ground and play dead when disturbed. Ground nesting birds freeze when approached. The hog-nosed snake and some birds appear to faint when handled. Ducks sometimes respond to the attack of a fox by death feigning. A group in Canada made a video of an event that provides insight into the adaptive significance of death feigning. A mallard duck was captured and mouthed by a red fox. The duck played dead while the fox dug a hole and buried it. The video sequence shows dirt contacting the cornea of the eye without the duck as much as blinking! As soon as the fox retreated, the duck flew to safety. It is obvious that death feigning saved the performer's life. Historical records of early hunting are replete with accounts of human hunters avoiding death by acting dead while being mauled by an angry bear. Even today we hear tales from the battlefield of solders surviving an overwhelming enemy attack by feigning death. It appears the human emotional faint represents our own body's response to strong emotion or threat by involuntary unconsciousness. The opossum would be the next subject of study, but I needed an assistant.

Geir Gabrielsen and I maintained an active correspondence since our meeting at the Oxford University telemetry conference. He completed his doctorate in physiology and continued to study the fear response of arctic animals. He worked with the Norwegian Polar Research Institute and directed an arctic research station on the Spitsbergen Islands. The field station was the most northerly permanent settlement on earth and represented an extremely harsh environment for man and beast. He described the passive fear response for every arctic animal he telemetered. He came to the United States to complete a post-doctoral study at Massachusetts Institute of Technology. Toward the end of his tenure he planned a trip to Oklahoma for a visit and for research. We would study an animal unfamiliar to him, the American Opossum.

Opossums are plentiful and easy to collect in western Oklahoma. They are often seen at night alongside roads searching for carrion, and may be caught by quickly grabbing their tails. They are omnivorous in the true sense of the word. Opossums eat berries, fruit, insects, worms, amphibians, reptiles, small birds and mammals, as well as road killed carrion. Like all other marsupials, the young are born highly immature. Gestation is only thirteen days, and the young at birth weigh less than an adult bumblebee. After making their way to the mother's pouch, they attach to her teat. In a month or so they may be seen peeking from their mother's pouch. Later they travel cross-country with their prehensile tails wrapped tightly around the mother's tail, which she keeps arched over her back. Although twelve or more may be born, usually only six to eight survive to weaning at the age of three months.

Unlike some of the other animals I had studied, there was a rich scientific literature regarding opossums. Death feigning had been the subject of several studies. Interestingly, the literature was equally divided regarding even the existence of the response. Some investigators, failing to elicit the reaction, dismissed "playing 'possum" as merely unsubstantiated American folklore. Others went to great length to build elaborate "mechanical dog jaws" to elicit the response. Of the scientists successfully eliciting the reaction, all concluded death feigning was an unconscious state—a behavioral faint. Few scientists had observed the response under natural conditions. Having spent many hours playing with 'possums as a boy, I knew without a doubt they played dead when attacked by a dog.

We secured our experimental animals by a combination of box traps and driving country roads late at night. The second morning we checked our traps resulted in one opossum and the accidental capture of a stripped skunk. I was not looking forward to releasing the unwanted pest, but I could hardly expect my guest from Norway to risk contamination. I was approaching the animal on my hands and knees with a long stick, in hopes of gently turning the trap over and releasing the varmint before it could spray me. From past experience I knew there was only about a 50:50 chance of my getting out of this ordeal smelling good.

Suddenly, Geir began shouting, "Stop, stop!"

I stopped. He asked, "Isn't that an American skunk?" I replied to the affirmative. He asked if it would spray him if he approached the cage. Again, I answered in the affirmative. His next response betrayed his dedication as a true teacher and animal behaviorist. He explained that he taught his students about the strong noxious odor of the American skunk, but had never actually smelled it. All his information was based on reading. He was excited about meeting (and yes, smelling) a live skunk at last. He moved toward the cage.

I lean strongly toward physiology: how living things work. Geir was a true animal behaviorist. I told him what to expect from the skunk. He observed the skunk's early warning of stamping its front feet firmly on the trap floor. He observed the skunk's final warning of stamping its feet while directing the base of its tail toward him. And yes, he observed the skunk's scent being released and sprayed on his clothes and in his face. He was ecstatic. He could share first hand (actually "first nose") knowledge with his students about the American skunk's behavior and pungency. He had to undress and bathe in the barn. He discarded the contaminated clothes. Science is such fun!

After just two nights of trapping and driving the roads, we collected four opossums. Transmitters were attached and a new study inaugurated. To make the experiments as natural as possible, the animals were released in an 80-acre pasture well after midnight. Red cedar trees provided cover for the investigators. The three-quarter moon yielded sufficient light for observation and measurements. As usual, each experiment began with thirty minutes of undisturbed control data.

Then, in an attempt to duplicate a predator's attack, the investigator approached the animal and made menacing noises. The animal was gently struck with a stick, which it promptly bit. A tug-of-war ensued. After one minute of anger response the animal was quickly grabbed, shaken and dropped. Instead of playing dead it ran away.

The experiment was repeated with each opossum. Every animal gave the same response. We tried pinching, scratching, and squeezing the animals. In each case they merely ran away upon release. It is easy to see why other scientists had given up and called the response "unsubstantiated American folklore." Having seen the response many times, I used the only device I knew would work. The services of Lady, my pet Border Collie, were again required for research.

Thirty minutes of undisturbed data were collected. Again, one minute of anger response was elicited. Lady was then commanded to "Sic 'em." Without hesitation, she grabbed the opossum and shook it. Upon being mouthed by the dog, the opossum immediately feigned death, and Lady dropped it. The opossum was limp, its tongue and lips deep blue from hypoxia. Heart rate and respiration dropped over 90%. It was unresponsive to touch, rough handling or pinching. Touching the cornea of the eye brought no response. All the evidence indicated the animal was unconscious. We, too, would have reached the same conclusion as the previous investigators had it not been for my four legged assistant, Lady.

After confirming that the animal was indeed playing dead, we retreated to measure the duration of the response. Although the animal remained motionless, its heart rate was slowly returning to pre-stimulus values. On her own, Lady re-approached the opossum. I yelled for her to stop and she did before even touching the animal; yet the opossum's heart rate plummeted. The meaning was obvious. In spite of total unresponsiveness to touch, the return of the dog deepened the bradycardia. The opossum was aware of the approach of the dog. It was fully conscious. Although we had no direct measurements of blood pressure, the fact that the animal was conscious strongly suggested blood pressure was maintained. Without the help of a dog the whole death feigning experiment would have failed. Had the dog not been disobedient and made the unplanned second approach, we would have accepted conclusions in the scientific literature that death feigning was an unconscious state. Now we knew better. Serendipity continued to play an important role in my research.

The experiments were repeated with the other opossums and identical results obtained. We had good data worthy of publication and I had several more pieces to the puzzle. The opossum experiments had many facets. Research is exciting because the unexpected often happens. Not only were the data rewarding in their own right, but also what happened to those data proved entertaining in unexpected ways. Two fascinating twists followed: one was a sad commentary on an extreme view by some uninformed people in our society, and the other was an exciting and unexpected application of these data to human illness.

Death feigning in an opossum

The publication of a technical paper is a long drawn out affair. It is not merely a matter of researching the related scientific literature, doing an experiment, writing about the results, drawing graphs, and sending it to a journal. Contemporary scientific journals practice what is called the peer review process. The manuscript and copies are submitted to an appropriate editor of a prestigious journal. The editor in turn sends copies of the manuscript to two or three other scientists knowledgeable in that particular area of research. They in turn read the manuscripts, make suggestions, and critically review the effort. Eventually the manuscripts are returned to the author. The reviewers' recommendations fall into three categories. First, the paper may be accepted as submitted. This is a rare event and happened to me only once. More commonly the paper is provisionally accepted. Changes must be made and the revision again reviewed. Often additional statistical tests must be performed and new graphs plotted. Finally, a paper may be rejected with no possibility of publication. It is then usually resubmitted to a less prestigious journal. Once a paper is accepted it often must wait in line for several months, or in some cases years, for the actual publication. Only researchers tend to appreciate this lengthy process. Experiments described in the most recent scientific journal are actually several months to a few years old. Information in textbooks is often five to ten years old. (This is changing because of the Internet, and it could have far reaching influences on how we do science.)

In writing a technical paper one must appease not only the editor, but several reviewers as well. The writing style is rigid and varies from journal to journal and from decade to decade. The proper literature must be cited. Statistical analysis is performed to meet the winds of current protocol. More often than not I completed the entire statistical analysis to please an editor. In my heart rate work, if the changes in heart rate evoked by the stimulus were not obvious, I was not motivated to publish the results. The use of statistics is akin to a drunk leaning on a street lamppost. The lamppost was intended for illumination, not support. Many reviewers seem to miss this important application of statistics. It is often used for support.

Even with the experience learned from over 100 technical publications, I was not prepared for what happened with the opossum data. Geir and I analyzed the data, prepared the graphs and researched the literature. We wrote the paper and were anxious to get it into print. We submitted the journal to the highly respected *American Journal of Physiology*. This is one of the most widely read medical journals and I had published in it several times previously. It was rejected, but not because of the poor data or weak analysis or sloppy graphs or failure to cite the proper literature. It was rejected because we used a dog to stimulate death feigning in the opossum. The editor feared pressure from animal rights activists. We had tried every other method available to elicit death feigning without success. What could be more natural than a predator like a dog attacking an opossum? There was no apparent injury to the animal; the skin was not even broken. At the end of our research each of the animals was released unharmed. Yet some people would think such an experiment was somehow unnecessarily cruel. So be it.

We revised the manuscript to the style of another journal and submitted it again. Again it was rejected for the same reason. Publication of the opossum data was delayed for over two years. We failed to find any American scientific journal willing to publish the data and were forced to publish it in a less widely read Norwegian journal. As we will see in a moment, that delay in publication may have cost human lives. Certainly experimental animals should be treated humanely, but I believe the value of human life exceeds the value of any animal.

Misinformation abounds among animal rights groups. They don't seem to understand that only happy, contented, well-cared-for animals will provide useful data. If an animal were sick or malnourished, the data obtained from such an animal would be worthless. If some people do not believe animals should be used in research that is fine with me. That is a personal decision and they certainly have a right to that opinion. But such people should be consistent in their actions. They must not seek medical help when they or their children or pets become injured or ill. The reason is obvious. All vaccines, all surgical procedures and over 80% of other medical practices in use today were developed using research animals, and the same is true for veterinarian procedures.

The only reason we use animals in research is because we have no alternatives. Computer models are not helpful until the information is first entered into the computer. Plant research and tissue culture experiments can not answer the myriad of pressing questions of medical research. Animals are necessary for research and will continue to be used as long as there is human and animal illness and injury. To do otherwise is inhumane.

The other twist taken by the opossum data was just as unexpected, but far more positive. My personal motivation to study animals is to know more about the animals. I am not motivated to do physiological research in order to relieve human suffering, cure disease or prolong life. I am motivated simply by a love for, and curiosity about, animals. I study animals because I want to know more about them.

It was Geir who made the connection to human illness. He returned to the east coast excited about our opossum data. We had taken several photographs during the experiments, and he showed a pediatrician a photo of an opossum feigning death. The lack of blood flow was evident from the bluish color of the lips and tongue. Geir was ecstatic because he and I suddenly had a deeper understanding about death feigning in American opossums than anyone else in the world. Together we had gotten a glimpse of yet another mountain. We were also developing a clear picture of the entire passive fear response. The pediatrician was ecstatic for a different reason. Upon seeing the photo and hearing Geir describe death feigning in opossums, the doctor saw a piece to a very different puzzle.

Sudden Infant Death Syndrome or SIDS is the largest single cause of death for infants from 6 weeks to 2 years of age. For no apparent reason they simply stop breathing and die. The breathing center in the back of the brain is thought to be involved. Unlike other infant deaths there is no obvious cause. Although the cause of death is lack of oxygen from failure to breathe, no one knows what triggers the fatal episode. Death often occurs at night while parent and baby sleep. Geir and the physician performed a simple experiment. They approached cribs where babies at risk for SIDS were sleeping. Geir clapped his hands loudly one time. A majority of the infants suddenly held their breaths, turned blue, and might have died had not the physician been nearby to resuscitate them. It appeared that under certain circumstances a baby at risk for SIDS might be literally frightened to death. Perceived fear might be a trigger. This was an important piece to the puzzle of SIDS—a piece discovered by accident in an 80-acre pasture under moonlight with an opossum and a disobedient dog.

Of course, there is still something else wrong with the child because fear should not cause an infant to hold his/her breath until he/she dies. There is a much larger picture related to SIDS, and other scientists and medical doctors continue to fit those pieces together. Perhaps in the near future this terrible disease will be eliminated. It is satisfying to know that my own research played a small part in understanding one small aspect of this tragic illness. And it started long ago at Big Lake when a telemetered alligator slowed its heart rate when it was frightened by the approach of another field zoologist in a canoe.

This illustrates how science advances, and also provides clear justification for continued support of basic science research. Basic research, unlike applied research, is not goal driven. A scientist is simply following his/her curiosity about the world we all share. No one knows where a discovery might lead. And it is this unknown that makes scientific research an exciting and rewarding pursuit.

# 13

# Tell me about the jungle

*Life is first boredom, then fear.*

—Philip Larkin

The sun was rising below the left wing tip of the Boeing 747. The captain announced, first in English then in Portuguese, "Good morning and welcome to Brazil. We have just crossed the equator." Weary travelers yawned and stretched. Some returned to sleep. I was elated. The window seat provided a panoramic view of the vast South American rain forest. The world below was green as far as the eye could see. Towers of smoke confirmed that the slash-and-burn method of clearing the forest was still being practiced. At last, I was going to see the jungle.

My earliest memories of the jungle were from stories told by missionaries returning from Africa. Two or three times each year missionaries from around the world visited our church and told of their work. I still recall their unusual displays of native artifacts and beautiful skins of impala, crocodile, zebra and python, and their stories about what seemed a fantasy world to this Oklahoma boy. After the service, people would stand in long lines to shake hands and ask specific questions about their ministry. Most would contribute. As a child, I would stand in line patiently, and when my turn came would simply say, "Tell me about the jungle."

As I looked out the airplane window, I thought of the jungle and the study on which I was about to embark. Dr. Sara Huggins, from the University of Houston, had been studying blood pressure in three-toed sloths in Recife, Brazil. Years earlier, I had designed a radio telemetry system that enabled her to monitor electroencephalograms (EEG) of mature alligators. Back then, she wanted to analyze brain waves to determine if alligators slept while basking in the sun. I helped install the device on a 700-pound alligator at the Rockefeller Wildlife Refuge. Although the results of that study were inconclusive, she and I stayed in touch. Years later she co-sponsored my becoming a member of the prestigious American

Physiological Society. Now she wanted to study the heart rate of sloths, and had asked for my assistance with the telemetry portion of the study. I was willing to make the sacrifice to visit a tropical rain forest.

Her contact in Brazil was Dr. Carlos DaCasta, Chairman of the Department of Physiology and Pharmacology at the Federal University of Pernambuco. Dr. DaCasta was one of the most interesting and congenial people I had ever met. Born in a Portuguese speaking province of India, he had a medical degree from India and earned a Ph.D. in physiology from the University of Houston medical school. Unlike most physiologists, he understood electronics and was himself an amateur radio operator.

It was in Houston that he and Dr. Huggins became acquainted. Upon completion of his training, he was offered a position in Brazil, due partly to his fluency in Portuguese. He accepted the position and soon became the department chairman. With the help of Dr. Huggins and Dr. DaCasta, I secured funds for travel and lodging from the U.S. National Science Foundation and the counterpart organization in Brazil for a site visit and preliminary study. After a brief layover in Rio de Janeiro, I was finally on my way to Recife on a small Brazilian airline. I was soon initiated to some of the profound differences that characterize developing nations. I was apparently the only Yankee on board. Only one of the flight attendants spoke English, and only a little. Once the plane was on automatic pilot, the captain left the co-pilot in control and came back and introduced himself. Since English is the international language of aircraft controllers, he had learned some English and wanted to practice with me. Trying to find some common ground, I fumbled in my wallet and produced my private pilot license. He was pleased to meet a fellow pilot and invited me up to the cockpit. After showing me some of the controls, he asked if I wanted to sit in the pilot's seat. Of course I jumped at the opportunity and was soon filled with self-importance, sitting at the controls of an old 727. He offered me some native fruit and a small cup of very strong coffee. I enjoyed both. Years later, while visiting my daughter in California, I tried some espresso. It tasted just like that Brazilian coffee. Perhaps Californians are not as advanced as they like to think! He asked about my visit to his country, and I tried to explain that I was a zoologist studying sloths. I think all he understood was that I was some sort of doctor. We chatted a while and I returned to my seat, happy to be so far from home and headed at long last to see a jungle.

As we descended for our first landing, I kept looking for the city. There was none. There was only a small village of perhaps 3,000 people. There wasn't even an airport control tower. The runway was short and rough. We hit hard, and due to the shortness of the runway, the pilot immediately applied the thrust reversers at full force. One of the flight attendants was assisting a passenger, and she was still standing when we landed. The hard landing and rapid deceleration caused her to fall forward, and she injured her ankle. As soon as we were stopped at the "ramp", the pilot came back and asked me to help the injured flight attendant. I could not make him understand that I was not a medical doctor. He insisted, saying we had two more stops before we would be in a city large enough to have a physician.

The flight attendant's ankle was already swollen and appeared badly sprained. She was clearly in pain and was weeping softly. All I could think of was to elevate her foot and pack it with ice to reduce further swelling. We also had to keep her warm. The other flight attendants helped and asked some passengers to move so she could lie across three seats. We applied ice and elevated her injured leg on a pile of pillows. We buried her with blankets, and I was able to get some aspirin from a passenger. She was still in pain and appeared very weak. I was worried about shock, but powerless to do more. I prayed.

After what seemed an eternity, we made two more landings at even smaller villages. It was difficult for me to understand cities large enough to have a commercial airport, yet not having a resident physician. At each of the small airports, something else was missing. There was no visible aircraft support equipment. There was no ground crew, no starter trailers; there was not even a tractor to push the aircraft away from the ramp. I was impressed to see the airplane's own thrust reverser used to back the airplane away from the ramp. It had never occurred to me that jet aircraft have a built-in reverse. I was reminded of our own affluence and waste. Perhaps aircraft unions also play a part in the difference. In the U.S. we are accustomed to seeing dozens of pieces of support equipment and at least that many people bustling around assisting arriving and departing commercial aircraft. Perhaps we could do with less. In route to the third city, the pilot informed the airport about the flight attendant's injury. Immediately upon arrival, the attendant was carried to a waiting ambulance and she would finally see a physician and receive proper treatment. As Paul Harvey often reminds us, "We do not all live in the same world."

Our approach to Recife was spectacular. The region was hilly. There were endless miles of sugar cane fields. The trees were distinctive and jungle-like. The towns and buildings looked different. It was obvious I was about to enter a world I had only dreamed about.

Recife is a beautiful city overlooking the azure Atlantic. It is the most eastern city in South America and lies closer to Africa than to the western side of South

America. Centuries ago, slave traders from Africa stopped at Recife for food and fresh water. Sick slaves were abandoned, but many survived and populated the region. Today, tourists from all over South America visit Recife to observe its unique culture. African influence is evident in the language, food, dance, dress and religion. Although there is an important tourist industry, no American tourists were seen and few people spoke English. Due to my fair skin and red beard, several Brazilians mistakenly thought me to be German and spoke German to me. My German was even worse than my Portuguese.

My hotel was near the beach, so I started each day with an invigorating swim. Joggers and swimmers alike used the beaches early in the morning. One morning I was swimming about 30 yards from shore and was caught in a strong rip tide. I am a strong, if not stylistic, swimmer, but it was all I could do to stay afloat. I was being pulled backward out to sea. Panic overtook me. I wanted to cry out for help, but did not know the Portuguese word. There were dozens of people on the beach, but I was becoming exhausted, and was perhaps 100 yards off shore and going backwards. And something was pulling me down.

The undertow subsided as quickly as it had started. I rolled over and floated to catch my breath, still moving rapidly out to sea. Soon the rip tide stopped as well. I swam about 20 yards parallel to the shoreline, then had an uneventful but lengthy swim back to the beach. Still visibly shaken when Carlos came to take me to the University, I described my adventure. He said I was lucky because just two weeks earlier two children had been caught in the same rip tide and both drowned. For the remainder of my visit, my morning exercise consisted of jogging on the beach.

An interesting event occurred each Saturday throughout this region of Brazil. People from the countryside poured into the town squares and set up booths. The event was called "hippie fair," so named because it was started in the 60's when hippies from the United States fled to Brazil to raise and smoke marijuana. They made handicrafts and brought them to town to sell each Saturday. Few hippies remained, but the tradition continued. Many of the booths contained homemade food; others sold souvenirs, some marketed clothes. Prices were unbelievably low. I purchased a handmade sailboat for less than five dollars, with the kind of detail that would bring hundreds of dollars at home. Beautiful carved wooden wall hangings could be purchased for less than a dollar. Hand woven baskets sold for pennies. It was an American shopper's dream, except for one problem—credit cards were not accepted!

The University was impressive. It contained the only medical school for hundreds of miles. Library support was adequate. The laboratories were staffed with highly trained and dedicated professionals. Most labs had some good equipment, but unfortunately much of it was inoperative. Grants were obtained to purchase the equipment, but once delivered, the ever-present federal red tape made it impossible to purchase replacement parts.

Electrical power was of poor quality. Frequency and voltage were unstable. Computers were particularly sensitive to power fluctuations, and were often unusable. Good hand held calculators could do much of the data analysis, but even they had to be smuggled into the country. They had a rigid rule, "If a product is made in Brazil, it can not be imported." The only Brazilian-made calculator was a cumbersome, unreliable four-function model. Batteries for electronic equipment were unavailable. Many high schools in the United States had better microscopes and other equipment than did the Brazilian universities and medical schools. People in our country panic with double-digit inflation, yet Brazil had triple-digit annual inflation for over ten years. In spite of the equipment problems, I met many highly motivated, competent scientists. They were doing respectable work and were publishing in some of the world's leading scientific journals.

Lunch time was my favorite event at the University. The professors each brought a small serving of something they had prepared at home. As noon approached, Dr. DaCasta would ask the technicians to go collect fruit. They would go out on the campus grounds and pick up fallen fresh fruit or throw sticks up into the trees, causing ripe fruit to fall. Those trees destroyed what little I thought I knew about plant hormones. One large branch contained fresh fruit, another branch on the same tree had green fruit, still other branches were flowering and some appeared dead. I was told this occurs all year. There were no seasons. We were less than 10 degrees south of the equator. After the fruit was collected, lunch was served. We ate together family style. Over half of what we ate was exotic tropical fruits collected on campus. Little meat was eaten, and the food was delicious. Even the local ice cream stores contained exotic tropical fruit flavors My favorite was guava.

But, I remembered, I was here to study a tropical animal.

When I was young and macho I wrestled alligators. Now that I was on the better side of 50 there was still an animal I could keep up with. Sloths are the slowest animals in the world. They spend their days hanging upside down from tree limbs. Their metabolism is a fraction of what one would expect and their body temperature is lower than other mammals. Most animals can flee when frightened. Not the sloth. Even the nerves that operate the muscles are slow acting. They totally lack red muscle. They simply can not hurry. Here is an animal for which fight or flight behavior is impossible. What happens to their heart rate when they become frightened? Sloths proved to be docile and easy to handle. When picked up they did not bite, scratch or try to escape.

Dr. Huggins had been investigating how sloths regulated their blood pressure and made a startling discovery. When approached by the investigator, heart rate dropped while blood pressure increased! My own earlier research with swamp rabbits suggested a simultaneous reduction in heart rate and muscle blood flow. Here was possibly an example of over-compensation. The sloth might provide an excellent model for future studies dealing with the relation between stress and hypertension or high blood pressure. My role was to attempt to monitor heart rate of free-ranging sloths. Since this was an internationally endangered specie, surgery was prohibited. Dr. Huggins designed a vest for the sloth to wear. The vest contained an inside pocket for the transmitter. Electrodes were pasted on, and the sloth was dressed in the vest. Everything checked out satisfactorily. It was time to head for the jungle.

Sloth, the world's slowest mammal

The area around Recife was mostly sugar cane fields. We drove nearly three hours to find a suitable place to release the sloth. It was on Dr. DaCosta's plantation. The region was all second growth, but it was beautiful. After releasing the sloth in a suitable forage tree, I asked if I could wonder around the forest. I was not concerned about venomous snakes or other wild animals, but expressed concern about poisonous plants. Dr. DaCosta said there were none. I disappeared into the jungle and quickly found a large lizard. I gave chase, and the first bush I ran into had two-inch poisonous thorns. Each puncture wound swelled and stung like a bee sting. I pondered the misinformation I had been given, and later asked Dr. Huggins about it. She laughed and said in Brazil, professional people are proud they don't have to hunt for food like peasants. Hunting and fishing were not sports of the educated, but necessities of the poor. Even though we were on Dr. DaCosta's own plantation, he had no first hand knowledge of the plants and animals that inhabited it. How different was the culture.

After my misadventure in the forest, I returned to the sloth, and we recorded heart rate data. As expected, the heart slowed when the investigators approached the animal. Here was an animal that could not flee, and it displayed the same response as other hiding animals. Unlike other animals studied, though, the fear response brought with it an increase in blood pressure. It is unfortunate that such a valuable research animal is threatened with extinction. Perhaps the lowly sloth holds the key to understanding another important human illness. Hypertension is a leading cause of heart disease and stroke and it is often correlated with emotional stress. The sloth is perhaps the best animal model in the world for hypertension research, but such studies are prohibited for this endangered species. We must learn that animals and plants have intrinsic value and can never be replaced. Like the bumper sticker says, "Extinction is forever." Exploitation of the rain forests affects all that inhabit planet earth.

# 14

# Blood pressure and the Timbu

*Though men pride themselves on their great deeds, they are often not the result of design but chance.*

—Duc Francois de La Rochefoucauld

There was another animal at the Medical School in Brazil that caught my attention. It was the Timbu or South American opossum. Several were caged in the animal room. I discovered they were left over from a previous study. Perhaps I could complete a second study during my ten-day visit. I obtained permission to attempt to get the animal to feign death. It was love at first sight. Timbu are delightful animals: slightly smaller than their northern cousins and more colorful, with a nearly white face and black ears. They have considerably more spirit than their American counterparts. When their cages were opened they often jumped aggressively at the investigator. I like animals with personality. Their behavior reminded me more of alligators or rattlesnakes than opossums. I tried to mimic my earlier study with American opossums. The experiment was begun with one minute of anger response. I growled at the Timbu and it growled back. I let it bite a stick and a tug of war heightened the anger. After the minute was up, I quickly grabbed the animal by the back of the neck and hip, shook it, and gently threw it down on a rubber mat. Without fail it played 'possum. Others were tested with the same response. I was ecstatic. I could induce death feigning in a more "scientific" way: without requiring the use of a dog.

Of even greater importance, I could finally measure blood pressure during death feigning. The mandatory use of the dog to elicit the response with the American opossum severely limited what could be done. Measurement of blood pressure when the dog was used was out of the question. Direct measure of blood

93

pressure requires cannulation of the carotid artery. Such a procedure would be easy in a medical school. All I needed was some expert help.

I was delighted to meet and talk at length with the other professor that was studying Timbu. Dr. Oldemir Ladowsky had been studying Timbu for several years and was interested in their embryonic development and nutrition. Marsupials provide zoologists with an excellent model for studying embryonic development, because much of the development that occurs inside the uterus of other mammals occurs inside the pouch with marsupials. I had hoped to find out more about their natural history and ecology, but once again I hit a cultural barrier. Educated scientists in Brazil seem uninterested in field studies. I wanted to know what animals prey on Timbu. The response was, "Nothing eats the Timbu for they are too smart." "Smart" is not one of the words I would have used to describe the Timbu or its northern cousin. Of course they had natural enemies, but such information was not available at the medical school, and I did not know the language well enough to talk to the locals who were familiar with the Timbu. I visited the university library to no avail. Because the Timbu is not considered to be an important animal and is of little economic value, not much has been published about its natural history. Sometimes we take for granted amazing things just because they have become familiar to us.

Dr. DaCosta selected three of his best medical students to assist me. They were Andre Louis de Andrade, Cicero Tiberio L De Almeida and Tania Marie Vireo. Each knew a little English and was competent. In Brazil, medical students do not attend college. They go directly from high school into a six-year medical training program. As is true in the United States, medical students have choices regarding certain training rotations. A research rotation was an option chosen by the three students that assisted in the surgical procedure. The procedure was simple and only required about 15 minutes. The animal was anesthetized and a small incision made to one side of the larynx. The external carotid was located, the small flexible cannula was inserted, and the incision closed. This enabled blood pressure to be monitored directly and recorded on a chart recorder.

Following recovery, the Timbu was connected to a blood pressure transducer and recorder and the earlier experiment repeated. In each case the animals performed as expected and feigned death immediately upon striking the rubber mat. Heart rate increased during aggressive behavior from about 200 to 275 bpm. Death feigning resulted in the heart rate dropping to 88 bpm, similar to the results of my earlier study with American opossums.

Aggressive Timbu

Of special interest was blood pressure. It dropped less than 10% during death feigning. This was significant because the old argument for such rapid slowing of the heart was always explained as a "carotid baroceptor response." The argument was that any sudden reduction of muscle blood flow would result in a transitory rise in blood pressure with a reflex slowing of the heart by the carotid pressure receptor. No such transitory hypertension occurred. This suggested that death feigning resulted in simultaneous reduction of heart rate and blood flow to skeletal muscles. Years later I found further support for this from a professor studying diving in rats in Hawaii. During forced diving he, too, found a sudden drop in heart rate without a transitory rise in blood pressure. It is nice to see patterns in nature. I think patterns are more obvious and certainly more rigid in physics and chemistry, but there are also fixed patterns in living organisms.

I had to travel nearly halfway around the world to get this important piece to the puzzle. I had no plans for studying Timbu. It was another accidental discovery. The picture of the passive response to fear was certainly becoming clearer than when I started, but it was far from complete. I had collected many pieces of the puzzle and a picture was finally emerging, but some of the details remained obscure. The search for the data had often been as exciting as the data itself. Zoological research is indeed an exciting adventure.

# 15

# Science, the unending quest

*Man masters nature not by force but by understanding. This is why science has succeeded where magic failed; because it has looked for no spell to cast on nature.*

—Jacob Bronowski

Someone outside science might think that thirty years of research on one problem would answer every possible question on the subject. Quite the opposite is true. Each time one question is answered ten others are raised. To me this is the most captivating aspect of science. A scientific quest is a modern variation of searching for the pot of gold at the end of the rainbow. The gold always appears to be just over the horizon. When that distance is traversed, the prize moves to the new horizon—always tempting yet always remaining just beyond reach. So it is with science. Scientists find themselves believing they will have the big picture if just one more experiment is performed or one more fact cataloged. But the big picture is as elusive as the pot of gold. No one ever has the final picture of nature, for nature is complex beyond comprehension. Yet we learn to find solace in small, incomplete pictures. Please allow me to share a small incomplete picture of passive fear with you.

Scientists have cataloged millions of facts and are searching for tens of millions of related facts. When those tens of millions of things are known there will be hundreds of millions of things yet to learn. Science is an unending quest, but the quest itself is rewarding. Discovery is tantalizing. Dr. Gabrielsen and I have been able to put a few of the pieces of the passive fear puzzle together. The studies we have published, along with some by other people, clearly show a myriad of physiological responses associated with passive fear. Table 1 provides a list of parameters now known to decrease during the passive fear response when the animal is attempting to hide.

## TABLE ONE

**Each of the following parameters decreases during passive fear response. Data from a variety of species.**

| | | |
|---|---|---|
| Behavioral activity | Heart rate | Respiration rate |
| Respiration depth | Metabolism | Oxygen consumption |
| Blood sugar | Brain blood flow | Body temperature |
| Heart blood flow | Skeletal muscle blood flow | |

Additional studies are needed to determine what happens to digestive and skin blood flow during passive fear and a host of other body functions, but this is a good start. It is obvious that passive fear has far-reaching behavioral and physiological implications. It alters virtually the entire animal and is every bit as important for survival as the active defense of fight or flight. Many additional studies must be done, but the importance of passive fear is now well documented and is beginning to appear in textbooks. Although I was the first to design experiments specifically to study passive fear, I will certainly not be the last.

Passive fear is also widespread in terms of the species displaying the response. Table 2 contains a current listing.

## TABLE TWO

**The following animals display the passive fear response with documented slowing of heart rate.**

**MAMMALS:**

| | | |
|---|---|---|
| White-tailed deer | Red deer | Deer mouse |
| Swamp rabbit | Cottontail rabbit | Eastern chipmunk |
| Ground squirrel | Woodchuck | Fox squirrel |
| Grey squirrel | Brazilian opossum | American opossum |
| Harp seal | Manatee | Man |

**BIRDS:**

| | | |
|---|---|---|
| Willow grouse | Svalbard ptarmigan | Pochard |
| Tufted duck | Common eider | Crested cormorant |
| Canadian goose | | |

**REPTILES:**

| | | |
|---|---|---|
| Ornate box turtle | Spectacled caiman | American alligator |
| Mountain Boomer | | |

**FISH:**

| | | |
|---|---|---|
| Bluegill sunfish | Chub salmon | Eel |
| Ganoid | Atlantic cod | Atlantic salmon |

Since it has been found in virtually every species thus far investigated, one would expect this list to continue to grow for decades. Comparisons must be made between free-ranging and captive animals. Laboratory studies only have real meaning as they approach parameters measured under natural conditions outdoors. With advances in modern radio telemetry this task will become easier and even commonplace. Some of the modern technologies have made my original studies child's play. Or looking at it the other way, it is exciting how much deeper today's technology will let us examine the lives of wild animals great and small.

Finally what do all these pieces show us? How do they fit together? Dr. Gabrielsen and I have come up with a first approximation shown in below.

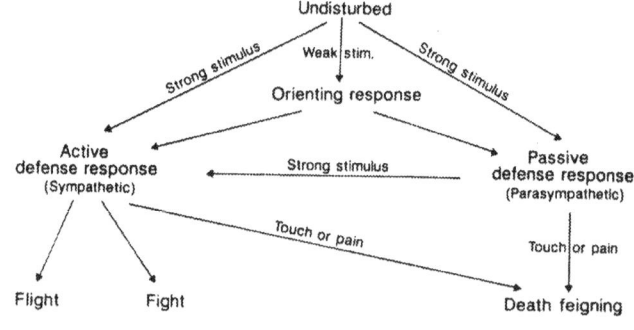

Active and passive fear response in contrast

No doubt this scheme will be modified as new data become available, but this was our attempt at fitting the pieces of the passive fear puzzle together. Notice that a stimulus to an undisturbed animal may elicit one of three responses. A mild stimulus usually results in the orienting response. Simply stated, the disturbed animal looks in the direction of the disturbance. In human terms it is collecting additional data to determine the cause of the disturbance. It is trying to identify the intruder as friend or foe or something that can be ignored.

Once the intruder is identified and found to pose a threat, either the active or passive defense may be elicited. Which response is used seems to depend on the availability of cover or some sort of hiding place. A rabbit may crouch low behind a rock or bunch of grass. A squirrel may hide on the opposite side of the tree. An alligator or seal may simply submerge. Near approach of the threat to the hiding animal usually results in conversion to active defense and flight.

Lacking a safe hiding place, the animal may choose the active response and run for cover. A woodchuck may retreat to its burrow. A squirrel on the ground may run to the nearest tree. A rabbit in the open may flee to cover. If cover is unavailable, the frightened animal may continue to flee. If cornered, it may fight. In an extreme case, touch or pain may elicit death feigning, as in the case of an opossum playing dead.

As I write this conclusion, I have mixed emotions about the fact that thirty years of research can be summarized in but a few paragraphs. The chase has been as much fun as the discovery. It is good that some pieces have fit together to make a picture, yet the picture is far from complete. I still hold a few pieces to the diving puzzle, yet do not have enough for a picture. I am convinced the classic diving response is derived from passive fear, yet do not understand why it is mediated differently. That mountain must await another explorer.

Perhaps one of the more important discoveries Dr. Gabrielsen and I made was the importance of studying wild animals outdoors under natural conditions. Wild animals certainly may be kept alive and healthy under laboratory conditions, but their behavioral and physiological responses to threat are clearly altered by confinement. Some animals failed to display the passive fear response under laboratory conditions. They lacked a necessary ingredient: a safe hiding place. This discovery will become more important as medical research begins to investigate more species of wild animals as important models to human illness. There is no doubt in my mind that this discovery would have eluded me had I trained in the more traditional laboratory environment. This explains why so few medical researchers understand the importance of outdoor field studies. Field zoology provides insight unavailable from textbooks or in the confines of a research laboratory.

And the converse discovery may be of even greater importance in years to come. Such a common and widespread response as passive fear could not be studied using traditional laboratory animals. Generations of selective breeding for docility have irreversibly altered behavioral and physiological response to threat. For such studies the medical researcher has no choice. He or she must recruit wild animals as experimental animals. I am afraid this approach will be even more difficult for many traditionally trained medical researchers. We are far too compartmentalized. Field zoologists and medical researchers attend different meetings, teach in different departments, read different journals and even drink coffee in different break areas. Their paths seldom cross. There is value in multi-disciplinary studies. I have learned far more from engineers and physicists than I have from other biologists. The reason is obvious. Other biologists and I were exposed to the same courses and acquired the same problem solving skills. It is the person outside our area that often brings insight and a new research approach. We must encourage cross- or multi-disciplinary courses and training. There is strength in diversity.

Another thirty years could be spent studying details of the fear response of just one specie. Nothing is known about the variability of the response among different individuals of the same specie. Just as there are no two identical sunsets, no two individual animals are exactly alike. There are differences due to genetics and differences due to learning, experience and age. Genetic differences of sexually reproducing animals are endless. Just as no two people look exactly alike, no two animals have the exact same genetic makeup. There are also learned differences. The old question of nature vs. nurture remains largely unsettled on many fronts. Learned differences are based on experience and no two individuals of the same specie have been exposed to the exact same experiences. Certainly heredity and experience modify the response of an individual to fear. What is frightful to one individual may be neutral or even positive for another. One is reminded of the Galapagos Islands where animals have no apparent fear of man due to lack of exposure (until recently).

Even within the same individual there are maturational changes of the fear response. Perhaps the best example is our own white-tailed deer. Fawns less than two weeks old freeze and slow their hearts when approached by man or coyotes, yet when the same individuals are just a week older they respond to the same stimulus by fleeing and accelerating heart rate. The age-dependent factor of the passive defense response needs to be studied for all animals exhibiting the response. These are obvious areas for expanded study. There are other less obvious, yet even more profitable areas for future study.

As scientific knowledge expands, the questions asked deepen. Most of my work was little more than cataloging what species respond passively to fear and how their heart rate changes. But the reduction in heart rate is only the tip of the iceberg. Heart rate is an algebraic sum of many factors. It is somewhat analogous to the speed of an automobile. The velocity of an automobile is related to many factors such as the condition and slope of the road, wind, the application of brakes or fuel, and the amount of horsepower, and perhaps even the mood of the driver at the time. So it is with passive fear. It is caused by many factors and in turn alters many factors within the body. No doubt the greatest studies of passive fear lie ahead.

As our knowledge of the passive response to fear in animals deepens, a clearer understanding of the human fear response will emerge. But there is more to science than facts and discoveries and breakthroughs. Scientific research has its own compensation. Doing the work of science is rewarding. Working outside with camera and binoculars while becoming one with nature is awe-inspiring. Discovering the secrets of how animals live and what they do and why they do it is the most satisfying thing I have ever accomplished. Animals do interesting things. Our respect for animals and all of nature increases as we try to fathom the complexities of even commonplace creatures. One of the most exciting aspects of scientific adventure is not knowing where it will lead. My curiosity about how alligators stayed warm started me on a journey of wonderment to how hiding animals respond to fear. And that journey lead to the crib of a baby at risk for an insidious killer. It is impossible to anticipate where future research into the passive fear response will lead. One fact is abundantly clear; it will be an exciting voyage into the unknown.

## THE END

# References and additional reading

Anderson, H. T. 1966. Physiological adaptations in diving. *Physiol. Rev.* **46:**212–243.

Arehart-Treichel, J. 1979. Monitoring the physiology of animals in the wild. *Science News* **116:**186–187.

Cannon, W. B. 1929. *Bodily changes in pain, hunger, fear and rage.* Second edition. Appleton and Co. New York and London.

Cannon. W. B. 1939. *The Wisdom of the body.* First revised edition. Norton and Company, New York.

Causby, L. A. and E. N. Smith, 1981. Control of fear bradycardia in the swamp rabbit, *Sylvilagus Aquaticus. **Comp. Biochem. Physiol.** **69C:**367–370.

Gabrielsen, G., J. Kanwisher and J. B. Steen, 1977. "Emotional" bradycardia: a telemetry study on incubating willow grouse, *Lagapus lagopus. **Acta. Physiol. Scand.** **100:**255–257.

Jacobsen, N. K. 1979 Alarm bradycardia in white-tailed deer fawns, *Odocoileus virginanus. **J. Mammal.** **60(2):**343–349.

Kennerly, T. E., Jr. 1974. Microenvironmental conditions of pocket gopher burrows. *Texas J. Sci.* **16:**395–441.

Scholander, P. F. 1940. Experimental investigations in respiratory function in diving mammals and birds. *Hvalrad. Sk.* **22:**1–131

Smith, E. N. 1974. Multichannel temperature and heart rate radio-telemetry transmitter. *J. Appl. Physiol.* **36(2):**252–255.

Smith, E. N. and W. E. Crowder. 1974. Implantable ECG transmitter employing a magnetic switch. *J. Appl. Physiol.* **36(5):**634–635.

Smith, E. N., R. D. Allison and W. E. Crowder. 1974. Bradycardia in a free ranging American alligator. *Copeia* **1974:**770–772.

Smith, E. N. and T. J. Salb. 1975. Multichannel subcarrier ECG, respiration and temperature biotelemetry system. *J. Appl. Physiol.* **39(2):**331–334.

Smith, E. N. and R. A. Woodruff. 1980. Fear bradycardia in free ranging woodchucks, *Marmonta monax. J. Mammal.* **61(4):**750–753.

Smith, E. N. 1980. Physiological radio telemetry of vertebrates. In *A Handbook on Biotelemetry and Radio Tracking,* (Ed. C. J. Amlaner and D. W. MacDonald) Pages: 45–55. Pergamon Press, Oxford.

Smith, E. N. and D. J. Sweet. 1980. Effect of atropine on the onset of fear bradycardia in Eastern Cottontail rabbits, *Sylvilagus floridanus.* **Comp. Biochem. Physiol. 66C:**239–241.

Smith, E.N., K. Sims and J. F. Vich. 1981. Oxygen consumption of frightened swamp Rabbits, *Sylvilagus aquaticus.* **Comp. Biochem. Physiol.** **70A:**533–536.

Smith, E. N., C. Johnson and K. J. Martin. 1981. Fear bradycardia in captive eastern chipmunk, *Tamias striatus.* **Comp. Biochem. Physiol.** **70A:**529–532.

Smith, E. N. and C. Johnson. 1984. Fear bradycardia in the eastern fox squirrel, *Sciurus niger* and eastern grey squirrel, *S. carolinesis.* **Comp. Biochem. Physiol. 78A:**409–411.

Smith, E. N. and M. C. DeCarvalho, Jr. 1985. Heart rate response to threat and diving in the ornate box turtle, *Terepene ornata.* **Physiol. Zool. 58(2):**236–241.

Smith, E. N., N. C. Long and J. Wood. 1986. Thermoregulation and evaporative water loss of green sea turtles, Chelonia mydas. J. Herp. **20(3):**325–332.

Smith, E. N. 1989. Biotelemetry Workshop—an intensive training session. *Biotelemetry X—Proceedings of the Tenth International Symposium on Biotelelmetry* (Ed. C. J. Amlaner, Jr.) Pages 462–477.

Smith, E. N. and Eric G. Aitken. 1989. Low power skin and muscle blood flow photo plethysmography biotelemetry system. ***Biotelemetry X—Proceedings of the Tenth International Symposium on Biotelelmetry*** (Ed. C. J. Amlaner, Jr.) Pages 326–331.

Smith, E. N. and Gary L. Barnes. 1989. Miniature low power blood flow photo plethsmography biotelemetry system. ***Biotelemetry X—Proceedings of the Tenth International Symposium on Biotelelmetry*** (Ed. C. J. Amlaner, Jr.) Pages 125–130.

Smith, E. N. and Steven E. Moore. 1989. Inexpensive magnetically switched temperature and ECG biotelemetry system. ***Biotelemetry X—Proceedings of the Tenth International Symposium on Biotelelmetry*** (Ed. C. J. Amlaner, Jr.) Pages 552–557.

# Meet the Author

Norbert Smith and a friend

E. Norbert Smith, Ph.D. was born in Oregon, but only remembers growing up on the family farm near Weatherford, Oklahoma where he was raised by his mother and maternal grandparents. As a child he loved animals and had many wild "pets" including especially snakes and lizards. In high school he became interested in electronics and earned his amateur ("ham") radio license. Five days after high school graduation he joined the Air Force with the promise of electronics school. After an honorable discharge from the service and working in electronics for three years, he returned to Weatherford to attend Southwestern Oklahoma State University majoring in Biology. While in college, he designed two radio telemetry systems. One was for Professor Hobart Landreth to track rattlesnakes. The other was created to monitor heart rate and body temperature of alligators. It was later used at Baylor University, where he earned a Master's Degree in Zoology studying alligator thermoregulation. He attended the University of California at Los Angeles two quarters and transferred to Texas Tech University where he graduated with a doctorate in Zoology and continued studying alligator thermoregulation. For over 30 years Dr. Smith has continued developing sophisticated radio telemetry systems and using them to investigate the behavior and heart rate response of a variety of wild animals to fear. Harvard Professor Edward Wilson, in his book *Naturalist*, said, "Science is modern civilization's highest achievement, but it has few heroes." Dr. Smith agrees. This book is about an average scientist's journey; his successes and his failures.

Dr. Smith has a varied writing background and has published numerous popular electronic magazine articles and nature stories for children. He has sold non-fiction material to a syndicated TV program and humor to *Reader's Digest*. He has published over 100 technical papers about the design of radio telemetry systems, reptilian thermoregulation, and the cardiovascular response of animals to fear. Most of his research was accomplished outdoors with free-ranging wild animals.

He has traveled extensively, lecturing at Oxford University in England and at universities and medical schools in North and South America. He has studied alligators in South Texas, sloths in the jungles of Brazil and sea turtles on Grand Cayman Island. His research has been showcased in *Science News* and on the *Today* TV show. The British Broadcasting Company included two segments of his alligator research in the widely distributed TV documentary, *A Smile for the Crocodile*. He has attached heart rate transmitters to more species of wild animals than anyone else in the world. He also has diverse academic experience, teaching medical physiology to optometry students, electronics to inmates of a federal prison and microbiology to nursing students, plus a variety of life science courses at universities in Kansas, New York, California and Oklahoma.

Following his divorce after 31 years of marriage, he needed more income, and started a new career as an over-the-road truck driver. It was an exciting ten year adventure and he drove over 2 million miles without an accident or traffic ticket. Always the teacher, he taught about 150 truck driving students how to safely operate big rigs. When he wasn't teaching, he practiced another hobby—photography—and took hundreds of roadside photographs. He recently retired from trucking and enjoys gardening, writing, nature photography, scuba diving and flying.

# Index

978-0-595-39096-0
0-595-39096-X

www.ingramcontent.com/pod-product-compliance
Lightning Source LLC
Chambersburg PA
CBHW051438280526
45785CB00003B/1335